The MACH-10 Leader

AI-Native Leadership at Decision Speed

Jason M. Riggs

HYPERSONIC
PUBLISHING

Published by Hypersonic Publishing.

ISBN: 979-8-9940323-5-0

First edition 2026

Contents

For everyone fighting to stay human in a world that demands more automation and less thinking.

For the builders, thinkers, and renegades who refuse to defend broken systems just because they feel safe.

Speed will reshape the system.

Judgment will decide what it becomes.

Soul will make it matter.

Preface

Why leadership has to change when time compresses

The MACH-10 PM® came from a simple reality: product managers were being asked to move faster, make better calls, and use AI without creating chaos.

That book focused on execution: how decisions get made, how work moves, and how product teams operate when speed isn't a nice advantage but the condition they're already working inside.

Since then, the pressure has only increased.

AI didn't create that pressure. It exposed what was already there and compressed the timeline around it. Work that took weeks now takes days. Work that took days now takes hours. Sometimes the outcome begins forming before the meeting ends, which is both impressive and mildly insulting.

Most teams now have access to the same tools and information. The advantage comes from how quickly an organization can turn signal into action.

Tools change. Physics doesn't.

That is decision speed. The MACH-10 PM examined what it means for product teams operating under that pressure. This book looks at what it means for leaders, because execution eventually stops being the only constraint.

Leadership becomes the bottleneck.

Decisions slow down, ownership gets blurry, and alignment begins replacing action. Everyone may be working hard, but the system still waits.

That failure usually doesn't happen because people are incapable. It happens because the way decisions are made no longer matches the speed of the environment around them.

The examples in this book come from patterns I've seen across product, business strategy, enterprise software, restaurant technology, AI automation, and large-scale technology environments.

The company names provide context, but the lesson is about the decision pattern, not the employer. Those patterns were shaped by work across environments including GoPro, Qualcomm, PAR Technology, and Audivi AI, where speed, judgment, customer behavior, and organizational complexity appeared in very different forms.

Across those environments, the stakes were practical: delayed pricing calls, compressed roadmap decisions, customer-impacting workflows, AI automation risk, and execution loops that either moved in time or lost value.

Some stories are drawn from direct experience, some are composites, and others are pattern-based. Where details could identify people, customers, or confidential decisions, I've generalized them without changing the leadership lesson.

This book isn't about one company, one team, or one tool stack. It's about what happens when time compresses, decisions carry greater consequence, and leadership has to evolve faster than the system around it.

The MACH-10 Leader™ is the leadership half of the model. The MACH-10 PM shows how to operate at speed inside a product function. The MACH-10 Leader shows how to create the conditions for that speed to scale across an organization.

You need both.

Execution without leadership alignment breaks down under pressure. Leadership without execution discipline creates direction without movement.

The MACH-10 Leadership Model™ follows a simple progression:

AI compresses time.
Compressed time shrinks decision windows.
Shrinking windows turn latency into liability.
Leaders must insert judgment earlier.
Judgment creates advantage only when it moves through faster learning loops.
Those loops endure only when ownership is clear.

That is the system this book is built around: compressed time, decision latency, judgment insertion, velocity loops, and accountable speed.

Intellectual Grounding

The MACH-10 Leadership Model is practitioner-built, but it connects to established work on decision cycles, organizational learning, automation risk, decision rights, trust calibration, and judgment under uncertainty. John Boyd's OODA loop, James March's work on organizational learning, Amy Edmondson's work on learning and psychological safety, Eric Ries's validated learning, Forsgren, Humble, and Kim's research on high-performing technology organizations, Lisanne Bainbridge's "Ironies of Automation," and research on automation bias and trust in automation all point toward the same underlying reality. Faster systems do not reduce the need for judgment. They make judgment more urgent. MACH-10 leadership brings those ideas into the AI-compressed operating environment, where the time between signal, decision, execution, and consequence keeps shrinking.

The model isn't meant to replace those bodies of work. It translates them into the operating pressure leaders are facing now. In a slower environment, decision cycles, learning loops, trust calibration, and ownership could be treated as separate management concerns. Under AI-compressed time, they collide. A delayed decision becomes value loss. A trusted output becomes a judgment problem. A slow feedback loop becomes competitive exposure. MACH-10 leadership is built for that collision point.

That's why this book uses the language of operating models instead of theory alone. The problem leaders face now isn't abstract. It shows up in missed windows, unclear decision rights, automation errors, slow feedback, and work that moves faster than judgment can follow. The research explains the pieces. MACH-10 leadership connects them inside the system leaders actually have to run.

The only question is whether you're shaping that movement or reacting to it after the fact.

That's why the work starts with the decisions in front of you, not with a transformation program that takes twelve months to explain.

How to Use This Book

This book is about leading when AI changes the speed of the system.

Read it as an operating model.

Don't read it as a theory to agree with later. Read it as a system to test now. The value of the book shows up when one decision moves faster, one loop gets shorter, one owner becomes clearer, or one AI workflow gets safer.

Each chapter is built around a leadership problem already showing up inside modern organizations: slow decisions, unclear ownership, alignment that becomes delay, automation that moves work instead of eliminating it, misplaced trust, and speed that creates motion without control.

Treat the ideas like tools. Put them to work.

Highlighting a sentence is nice. Changing a decision is better.

Start where your organization is already feeling pressure, whether it's a stalled decision, a bloated approval path, a team moving fast but creating rework, an AI workflow that creates cleanup somewhere else, or a strategy process that produces more alignment than movement.

Then use the chapter-end sections as diagnostics. Find where the system is slowing down, where judgment needs to enter earlier, and where ownership needs to become clearer.

You don't need to fix the whole organization at once.

Start with one decision.

Shorten the loop.

Name the owner.

Insert judgment where it matters.

Move while the decision still has value.

That's where MACH-10 leadership begins.

Chapter 1

Leadership Is Already Broken

The System Is Moving Faster Than Your Decisions

Something has changed, and most teams feel it before they can explain it.

It doesn't arrive as a clean break. It shows up as pressure. Decisions compress. Opportunities close faster. Work that once took weeks now takes days, sometimes hours. The world keeps moving while everyone is still talking.

You see it in the meeting that should've moved something forward but didn't, the decision everyone circled but no one made, or the moment when the answer was clear and it still took another week to act. The process may have worked exactly as designed. The problem is that the opportunity didn't wait.

That's what makes delay deceptive. Another review sounds reasonable. Another stakeholder conversation feels responsible. Another request for data looks like diligence. The cost only becomes obvious after the moment has passed.

The instinct is predictable: push harder and ask people to move faster inside the same operating model. I've done the same thing, and in a slower environment, it worked.

It doesn't anymore.

Most teams still gather input, align stakeholders, pressure-test assumptions, and wait for enough confidence to move. For a long time, that signaled strong leadership. But it depended on one assumption: that the decision window would stay open long enough to finish thinking.

That assumption is breaking.

Once time compresses, every review, handoff, and extra day consumes some of the value you're trying to protect. A careful answer can be completely defensible and still arrive too late to matter.

Caution still matters. Some decisions deserve more time. But if it takes three weeks to improve a decision by 10 percent while the opportunity loses 30 percent of its value, the process didn't improve the outcome.

It made the answer better and the decision worse.

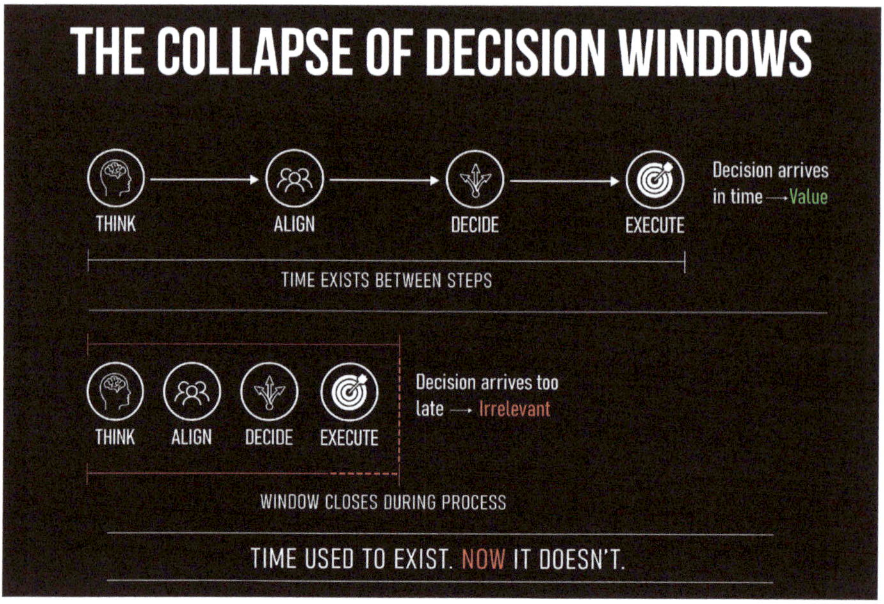

FIG 1.1 – Collapse of Decision Windows. The window used to stay open. Now it closes while you're still aligning.

I know what this looks like when it breaks.

We were working through a decision that mattered. Not a small operating call, but something that would shape where we invested, how we positioned our products, and what we pushed into the market over the next two quarters. It was the kind of decision you don't rush. We had the right people in the room: smart operators with good context and strong opinions. Nobody was phoning it in. We pulled the data, pressure-tested assumptions, looked at the trade-offs, and every time we got close, we slowed down just enough to make sure we were getting it right.

It felt responsible. It felt like leadership.

It also took time.

While we were doing that, the market didn't pause. Competitors were moving without perfect information, but they were learning in real time while we were still refining the call. They were getting feedback from the market while we were still getting feedback from each other. That difference looked small in the moment, but it compounded fast.

By the time we landed it, the answer was solid.

It just didn't matter anymore.

The conditions had shifted, and the window we thought we were operating inside was gone. We executed well against something that was already behind by the time it hit the ground. Nothing about the choice was obviously wrong. We had simply spent too much of the decision's value getting comfortable with it.

It was just late.

And late created its own workload. The missed move was only the first cost. Then came the recovery work: re-explaining the decision, rebuilding momentum, and adjusting to conditions that had changed. What should've been a forward move became a catch-up effort. **We were now spending time fixing the consequences of having spent too much time deciding.**

That's how decision latency hides. It doesn't always look like failure in the room. Sometimes it looks like a very good meeting.

The bill arrives later.

YOU CAN BE RIGHT AND STILL LOSE

Being Right Is Not the Same as Winning

That's the part most teams learn the hard way.

You can follow the old playbook almost perfectly. Gather the data, pressure-test the assumptions, bring the right people into the room, and build a recommendation that looks solid from every angle. The analysis can be strong, the logic can hold up, and the decision can be completely defensible.

And you can still lose.

Sometimes the moment simply outruns the process. By the time the organization reaches the right answer, the conditions that made the answer valuable have already changed. That's the shift this book is built around.

One team's still trying to get aligned while another is already in the market learning from what happens. The first may have better analysis. The second has reality: customer reactions, operational feedback, and evidence about what actually happens once the decision hits the world.

The real constraint is decision flow: how quickly a signal becomes a call, and how quickly that call becomes action. Every handoff, review, approval, and unnecessary participant adds time. In a fast environment, that delay can consume the value of the decision itself. The goal isn't reckless speed. It's building a system that can learn while the opportunity still exists.

Sales makes this painfully obvious. A deal starts moving, the customer's engaged, and then comes another forecast review, another pricing discussion, another round of internal agreement. None of those steps looks especially dangerous on its own, but the customer experiences the accumulated delay. You used to have more margin for that. Now you don't.

A competitor gets access. The customer's priorities shift. The internal champion loses urgency. Eventually your team makes the decision, but now it's responding to a situation somebody else has already shaped. Outside the company, none of the internal logic matters. The customer only knows how long it took you to respond.

Operations has its own version: unresolved constraints, workarounds, added labor, and temporary fixes that quietly become permanent process. People teams do it too through slow hiring, delayed performance calls, and roles everybody knows need to change. Different function, same physics. The decision may wait, but the consequences don't.

And those consequences rarely wait politely.

A temporary workaround becomes the new normal. A known performance issue gets tolerated long enough that everyone else adjusts around it. By the time the decision finally gets made, the organization isn't solving the original problem anymore. It's solving the secondary problems created by waiting.

AI is accelerating all of this. It removes friction from execution, makes output cheaper, and compresses the distance between an idea and its consequences. Most organizations were built for a different speed.

That mismatch is becoming harder to hide.

When execution was slower, leaders could afford more space between observation and action. Research took longer. Analysis took longer. The entire system moved at a pace that gave leadership process room to breathe. AI changes that equation because the work downstream of the decision can now happen much faster than the decision itself.

Leadership models built around a comfortable gap between thinking and doing start to fail when that gap disappears. More information doesn't help if it arrives after the window to act. Alignment doesn't help if achieving it takes longer than the opportunity survives. Waiting doesn't reduce risk if the environment keeps changing while the organization stands still.

Most teams can feel this before they can name it. Decisions take too long. People are busy, calendars are packed, and somehow the things that matter most still aren't moving. Leaders respond with more process, more reviews, more visibility, and more people in the room because those are the tools that used to create control.

That response makes sense. It also makes the problem worse.

The organization thinks it's becoming more disciplined while adding weight to the exact parts of the operating model that need to move. Every new layer creates another place where ownership can blur and a decision can wait.

Eventually, delay starts calling itself alignment. Another meeting, another readout, another round of context. From inside the organization, it rarely looks dysfunctional. It looks responsible. It looks like leadership.

That's why it survives.

Nobody sets out to build a slow organization. It happens one reasonable decision at a time. One more stakeholder gets added because their input could be useful. One more review appears because the last decision went badly. One more approval gets inserted because nobody wants to be surprised. Eventually the operating model becomes full of perfectly understandable steps that collectively make it almost impossible to move quickly.

This is the alignment trap.

The room gets calmer while the opportunity moves farther away.

Getting everyone on the same page is supposed to reduce risk. In a slower environment, it often does. In a fast one, complete agreement can delay a decision past the point where agreement has any value. Coordination isn't progress, and an organization can look careful and well managed while quietly losing the ability to move.

Mindset Shift to Adopt

Speed is now the environment, not a luxury. The real test is whether your decisions happen while they still matter.

Tools to Try This Week

Track one decision from first discussion to action. Note every review, handoff, and request for more input. Identify which steps improved the decision and which only delayed it.

For example, a pricing decision may have nine participants, three reviews, and no clear owner. Reduce the group to the people with essential context, name one decision owner, and set a deadline based on when the decision loses value. The goal is to separate the steps that improve judgment from the ones that only create comfort.

Do This Now

Choose one decision that has been discussed more than twice without resolution.

Name the owner. One person makes the call.

Shrink the room. Remove anyone who doesn't add information that could change the decision.

Set the deadline. Make the decision within 24 hours, then communicate it and move.

Key Takeaways

The system is already moving faster than most leadership models can handle.

Decision speed now determines impact, not just decision quality.

Alignment delays are never neutral. They destroy momentum.

MACH-10 Leadership Moves

Replace one sync meeting with a decision deadline.

Move one decision before it feels fully resolved.

Create enough clarity to act, then learn from what happens.

The system is already moving.
The question is whether your decisions can keep up.

THE SYSTEM IS NOT SLOW ANYMORE YOU ARE

Chapter 2

The Collapse of Traditional Leadership

When Good Decisions Start Arriving Too Late

Chapter 1 showed the visible failure. Decisions were still being made, but they were arriving after the moment had passed.

Decisions don't die in meetings. They die in the machinery that makes those meetings necessary. This chapter is about the leadership model behind that failure.

Traditional leadership didn't break because it was foolish. It broke because it was built for a slower environment. More input, more alignment, and more process once gave leaders room to reduce risk before committing. Those practices created stability, but they also depended on something most organizations barely had to think about: time.

That time is disappearing.

When the space between decision and consequence collapses, the behaviors that once protected the organization can start working against it. Alignment can still be useful. Validation can still matter. Confidence still has a role. But none of them create value if they arrive after the window to act has closed. At that point, you haven't reduced risk. You've just traded the risk of moving too soon for the risk of moving too late.

I saw this firsthand at Qualcomm. We spent several weeks aligning around a fleet dashboard while the market window kept narrowing. The work was sound, the decision was defensible, and nobody was asleep at the wheel.

We were right. We were late.

I've seen the same pattern in restaurant technology, where a rollout sat for weeks waiting for alignment while operators absorbed the cost in the field. The technology worked. The organization simply took too long to get comfortable enough to move, and the people closest to the problem kept paying for that delay.

That's the real cost of waiting.

It rarely shows up as some dramatic failure in the moment. It shows up later as recovery work, lost momentum, and people paying interest on a decision that should've already moved.

This is where the old model breaks.

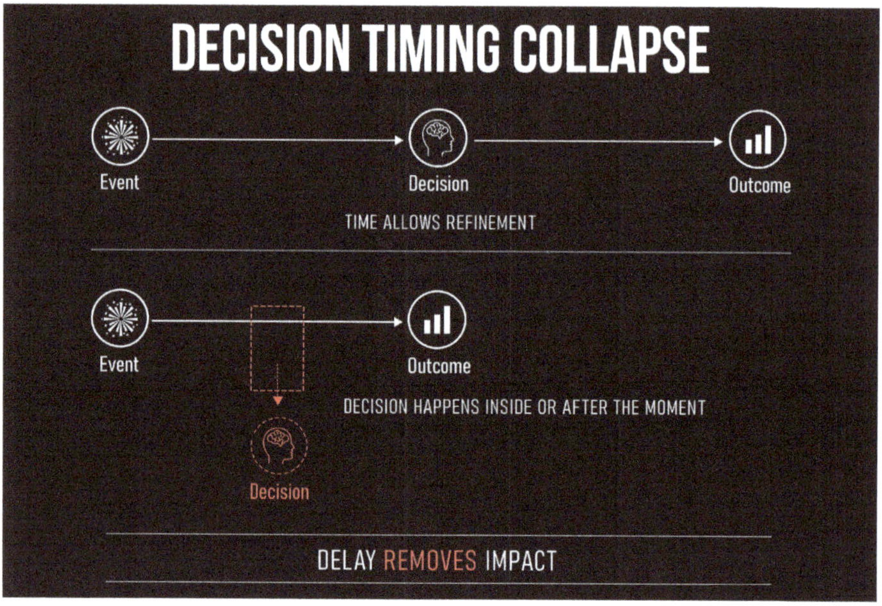

FIG 2.1 – Decision Timing Collapse. Delay removes impact when decisions happen inside or after the moment.

When Alignment Becomes Delay

When time compresses, alignment changes function. It can still improve a decision, but it can also become a way to postpone ownership while preserving the appearance of control. That's what makes the alignment trap so dangerous: from inside the organization, it rarely looks reckless or dysfunctional. It looks disciplined, mature, and responsible. More people get invited into the room, more context gets shared, and more time gets spent making sure everyone feels comfortable before anything moves.

Decision-rights research has made a version of this point for years. When authority is unclear, organizations compensate by adding coordination in the name of safety. In a compressed environment, every additional layer has to earn its place. If it improves the decision, keep it. If it only makes the organization feel safer, it's slowing the response at exactly the moment the organization needs to act.

This is where instinct starts working against the leader.

The natural response to uncertainty is to tighten the process. Leaders add checkpoints, pull in more stakeholders, ask for another round of analysis, and create one more opportunity to make sure nothing important has been missed. It feels like the responsible way to move the initiative forward.

It often does the opposite.

The organization becomes less responsive while convincing itself that it has become more careful. The playbook that once created consistency starts creating resistance because it was built for a different speed. A decision that arrives too late isn't much of a decision at all. It's evidence that the organization moved too slowly.

The process may have looked careful.

The outcome didn't care.

Leadership isn't defined by how carefully a decision is made in isolation. It's defined by whether that decision moves through the organization in time to matter. That doesn't mean stripping judgment out of the process. It means applying judgment earlier, knowing which questions could actually change the call, and recognizing which ones are just giving everyone another reason not to own it yet.

You don't need perfect information. You need enough clarity to act before the window closes.

Once the moment passes, the analysis may still hold up. The logic may still be sound. The recommendation may still be defensible.

The impact is still gone.

Speed isn't the enemy of quality. Delay dressed as diligence is.

The leaders who win now cut the wait, not the thought. They move while the decision still matters.

The Hidden Expense of Getting Comfortable

That's the part leaders have to learn to see earlier.

The delay isn't always inside the decision itself. Sometimes the answer is already visible, the trade-offs are known, the risks are understood, and the person who owns the call already knows what needs to happen. The organization just isn't comfortable yet.

Comfort isn't the same as clarity. Clarity tells the team what's known, what's uncertain, which consequence matters most, and who owns the decision. Comfort asks for more time because movement still feels exposed.

A useful review can surface a risk that changes the decision. A weak review repeats what everyone already knows. A useful stakeholder brings missing context. An unnecessary stakeholder makes the room feel safer without improving the answer. A useful analysis reveals a better path. A defensive analysis gives everyone another spreadsheet to hide behind.

That's harder than it sounds because most organizations reward visible caution. The person asking for one more review often looks responsible, while the person pushing the decision forward can look impatient or reckless. In a slower environment, that bias was easier to absorb because the window usually stayed open long enough for the process to finish.

A decision can lose value while the organization is still improving the ceremony around it. The meeting gets sharper, the summary gets cleaner, and everyone gets progressively more comfortable with the call while the moment it was supposed to affect keeps moving.

Each additional step can look reasonable on its own. Another review may cost only a day. Another stakeholder may add only one meeting. Another analysis may feel like a small price for more confidence. But the organization experiences the accumulated delay.

The question isn't whether every review adds some value. Most do. It's whether that value is greater than what the decision is losing while it waits. Once the same risks, trade-offs, and objections start repeating, the process isn't improving the decision anymore. It's delaying the moment when someone has to own it.

By the time the organization finally acts, it may have a better explanation and a weaker outcome.

That isn't discipline. It's comfort buying time the decision can't afford.

Signals from the Edge

I've sat in meetings where the useful thinking was finished long before the process ended.

The team had the right inputs, the trade-offs were understood, and the direction was clear enough to move. But the remaining discussion didn't change the answer. It only made the organization more comfortable accepting it.

That's where traditional leadership becomes difficult to see clearly. From the outside, the process looks healthy. People are engaged, risks are being discussed, and leaders are asking reasonable questions.

But the decision has stopped improving. It's simply waiting for permission, and the meeting has become a very expensive waiting room. Everyone is still contributing, but mostly to the furniture.

The problem is that participation can keep expanding even after judgment has stopped improving. Each person adds another perspective, another caveat, or another condition for comfort. The discussion becomes richer, but the decision becomes no clearer.

That's the point when leaders have to separate useful dissent from repeated hesitation. New information should reopen a decision. Familiar discomfort shouldn't. Once the same concerns are circling without changing the call, the process has shifted from analysis to delay.

Smaller teams often draw the line earlier. They define what they need to know, accept that the first action will be imperfect, and put the decision into the world while it still has time to matter. Within a week, they can have real feedback, a second iteration in motion, and a clearer understanding of what actually mattered.

The difference is discipline, not recklessness. They understand that certainty has an expiration date, and their advantage isn't capability. It's knowing when comfort has stopped improving quality.

Mindset Shift to Adopt

Stop treating alignment as proof that a decision is better.

Start asking whether the process is improving the decision or only making it later.

Tools to Try This Week

Audit one decision moving through the old leadership model. List every review, approval, handoff, and request for more information, then mark whether each step changed the answer or only delayed ownership.

For example, if a rollout required four reviews after the risks and owner were already clear, keep the review that improved the decision and remove or time-box the rest.

Use AI to summarize the available context, surface trade-offs, and clarify what's known. Don't use it to create another reason to wait.

Write down the earliest point when the team had enough clarity to move. Compare that moment with when action actually began, and treat the gap as decision latency.

Set a clear decision threshold, name the owner, and act once that threshold is met. Then track where the decision continued waiting after the answer was already clear.

Do This Now

Review three recent decisions. Write down how long each one took from first discussion to action.

Find the moment of clarity. For each decision, identify when the answer was probably clear enough to make.

Compare clarity to action. Look at when the decision could have moved versus when it actually moved.

Remove one delay. Cut or time-box one step that added confidence but didn't change the answer.

Name the owner. If a decision has no clear owner, assign one immediately. If everyone owns it, no one does.

Key Takeaways

Traditional leadership was built for a slower environment.

Alignment can create delay instead of clarity.

More information doesn't help if it arrives after the window to act has closed.

Decision quality and decision timing can no longer be separated.

MACH-10 Leadership Moves

Audit one approval path. Identify the step that adds comfort but doesn't change the decision.

Set a decision threshold. Define what must be true before the discussion ends.

End one alignment loop. Once the threshold is met, stop debating and move the decision forward.

Move with enough clarity. Don't wait until everyone feels fully comfortable.

Measure decision time. Track the time from first discussion to actual action.

A late decision is not a better decision.
It is a postmortem with nicer formatting.

**A DECISION
THAT ARRIVES
TOO LATE
IS NOT A
DECISION
IT IS A
POSTMORTEM**

Chapter 3

Speed Alone Is a Liability

Why Judgment Is the Only Thing That Makes Speed Useful

Once leaders see the environment moving faster, the instinct is obvious: accelerate everything. Move faster. Ship sooner. Decide more quickly.

But "move faster" isn't a strategy.

It's a volume setting.

Speed has become the default advice because every organization feels the pressure. Cycles are shorter, expectations are higher, and AI has made output cheaper, faster, and easier to produce. So the natural response is to push harder across the board: increase output, shorten timelines, remove steps, and compress decisions.

At first, it looks like progress. Work moves faster. Decisions happen sooner. Teams feel more productive.

Then the cracks appear.

Priorities collide. Teams move faster in different directions. Rework starts piling up. Decisions get made without enough context, and small mistakes compound into larger ones. The organization is moving fast, but that doesn't necessarily mean it's moving toward anything coherent.

That's the problem with speed. It doesn't create clarity. It exposes whether clarity was there in the first place.

If the operating model is aligned, speed amplifies it. If the organization is confused, speed amplifies that too. Speed reveals the structure carrying the work.

That's where teams get into trouble.

They hear that speed matters, so they start removing anything that slows movement: fewer reviews, shorter cycles, faster calls, less discussion, more output.

Some of that may be exactly right. But without judgment, speed becomes indiscriminate.

Automation research has been warning about a version of this for years. Faster systems don't remove the need for human judgment. They move that judgment to more critical points in the work.

The cost usually shows up later. The team ships sooner, but cleanup starts stacking up behind the scenes. Decisions have to be revisited, work has to be re-explained, and time saved in one place gets spent somewhere worse.

Speed without clarity creates chaos.

The answer is targeted speed: knowing which decisions need urgency, which need restraint, and which need just enough movement to create the next piece of truth.

That's where Speed with Soul™ comes in.

It's a simple idea, or at least it sounds simple until you try to run it inside a real organization filled with humans, calendars, politics, dependencies, and the occasional meeting that somehow exists only because another meeting happened first.

Speed means moving while the decision can still change the outcome. It keeps work from getting trapped in delay and process that makes everyone feel productive while nothing important actually moves.

Soul is the judgment inside that movement. It's context, prioritization, restraint, and the ability to look at incomplete information without pretending the picture is complete. It's knowing when to push, when to pause, and when the organization is simply producing more noise with better formatting.

Without speed, judgment arrives too late.

Without soul, speed starts breaking things.

You need both because pressure to act faster doesn't automatically make an organization smarter. In plenty of cases, it just gives weak assumptions a better engine.

Speed without soul creates bad decisions quickly. Soul without speed creates good decisions that arrive too late. Neither creates advantage on its own because timing is part of quality.

That's the part leaders have to own.

They make the call under pressure and act before the picture is complete because it almost never is. They accept uncertainty, move anyway, and stay alert for the difference between speed that's creating learning and speed that's simply creating motion.

That judgment shows up in how the work gets structured. Priorities are clear enough that people can act without waiting for permission on every move. Decision groups stay small. Feedback loops stay short. Real-world input replaces endless internal debate, and correction is expected instead of treated like evidence that somebody screwed up.

The best operating models don't make every decision perfect.

They make decisions easier to aim, easier to correct, and harder to hide from once reality starts talking back.

That distinction matters because speed changes what leadership has to protect. In a slower environment, leaders could often protect quality by extending the process: add another review, ask for another readout, wait for more confidence. The extra time felt like discipline because the window was still open.

In a faster environment, that logic starts to fail.

Leaders protect quality by making decisions easier to correct once they meet reality. That means being clear about what must be true, what remains uncertain, who owns the call, and how quickly the team will know whether the direction is working.

Speed with Soul doesn't ask leaders to pretend they know more than they do. It asks them to move with enough clarity to learn before the moment disappears.

Imperfection is allowed because waiting for perfection isn't.

That's why Speed with Soul matters. It gives leaders a way to separate productive urgency from performative motion. One creates learning, clarity, and momentum. The other creates noise, rework, and exhaustion dressed up as progress.

It's an operating model, not a slogan.

Speed is how fast you move. Soul is how well you understand what matters.

Advantage is created where those two intersect.

That's where speed stops being motion and starts becoming leadership.

FIG 3.1 – Speed with Soul Operating Model. Speed creates movement. Soul gives that movement judgment, context, and direction.

Most organizations increase speed before they strengthen judgment. That's why things break.

They accelerate with less clarity, and confusion scales right along with the output. More activity. More decisions. More work hitting the system faster. But if people don't know which calls matter, who owns them, or what can safely move without another layer of review, the organization just gets better at creating noise.

Speed creates value when it's aimed at the choices where movement can still change the outcome. Some decisions should move immediately. Others need tighter control. The difference comes down to reversibility, impact, and consequence.

Startups can often move faster because the blast radius is smaller. If the call is wrong, the organization can usually absorb the hit, learn, and adjust. That freedom matters because the cost of being late can be greater than the cost of being imperfect.

Large organizations face a different problem. One decision can affect revenue, compliance, customer trust, brand reputation, and how the market interprets the company. In regulated or high-consequence environments, leaders have to know which calls deserve more control and which ones are being slowed by inherited caution that nobody has bothered to question in years.

Some mistakes cost more and are harder to unwind. That still doesn't justify slowing everything down by default.

Most teams get this wrong in one of two ways. They slow everything down in the name of risk management, or they accelerate everything in the name of urgency. Both responses are easy to explain in a meeting. Neither is especially useful once the organization starts paying for the consequences.

The better question is, "Where does speed create advantage?"

The equally important question is, "Where does speed create unnecessary risk?"

That's targeted speed.

High-risk decisions shouldn't slow the entire organization, and low-risk decisions shouldn't inherit the process weight of irreversible ones. A small experiment shouldn't need the same machinery as a decision that can materially damage the business. Yet organizations do this all the time because process tends to spread once it gets established, whether the risk profile justifies it or not.

You can see the effect everywhere. In finance, slow budgeting decisions can stall every function downstream. In product, low-risk experiments can spend so long waiting for approval that the learning arrives after anyone cares. In customer-facing teams, slow recovery decisions can quietly erode trust while everyone is still debating the correct response.

Leaders have to know where speed belongs.

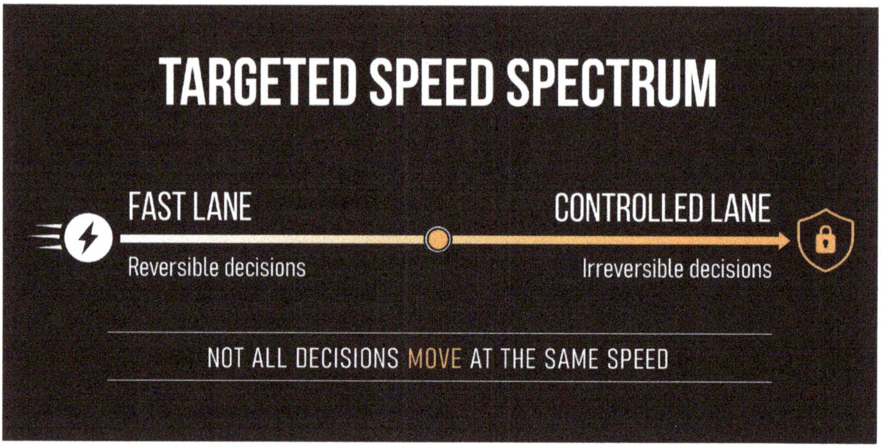

FIG 3.2 – Targeted Speed Spectrum. Not all decisions should move at the same speed.

That's the point of the spectrum.

Speed is a leadership choice, and the work is matching pace to consequence. A low-risk experiment shouldn't inherit the approval path of a major strategic commitment. A reversible product test shouldn't move through the same machinery as a pricing change, compliance decision, or customer trust issue. Once leaders stop distinguishing among those decisions, process spreads whether the risk justifies it or not.

When every decision carries the weight of the highest-risk decision, the whole organization slows. When everything moves like the lowest-risk decision, the organization creates exposure it can't absorb. Both are failures of judgment, just in opposite directions.

Soul is the judgment that tells the organization where speed belongs, where restraint belongs, and when the cost of being wrong is too high to confuse movement with progress. Leaders who get this right start with consequence rather than letting urgency set the pace.

If the decision is reversible, act quickly. Learn from contact with reality, correct as needed, and don't bury a low-risk call under a process designed for existential risk.

If the decision is hard to reverse, slow down enough to understand what can happen if you're wrong. Name the assumptions, identify the owner, define what must be true, and then move with intention. The goal is to apply control where it earns its cost rather than smearing it across every decision.

That's the line leaders have to hold: move quickly where learning is cheap and add control where consequences are expensive.

The Speed Law

Speed is not the highest value.

Speed yields to consequence.

If a decision is reversible, execute faster. If the blast radius is small, shorten the path. If the signal is fading, act while the decision still has value.

But speed has limits.

Speed yields to safety.

Speed yields to law.

Speed yields to ethics.

Speed yields to customer trust.

Speed yields when the consequence is irreversible or the blast radius is unclear.

Every decision has a speed limit set by consequence.

Speed where learning is cheap. Judgment where consequences are expensive.

That's disciplined control: speed guided by judgment.

The danger starts when the organization stops making those distinctions. Everything begins to feel urgent, actionable, and complete enough to use. Movement gets rewarded before understanding, and a system built to increase velocity starts creating risk faster than anyone can see it accumulating.

Leaders have to manage both sides. They can't slow the organization by default, and they can't accelerate it blindly. They have to decide where speed creates advantage, where it needs a guardrail, and make those choices before everyone is staring at the clock.

Because pressure is a terrible time to invent the rules.

Without them, urgency makes the decision for you. Teams fall back on whatever feels fastest in the moment, even when that pushes risk somewhere else. One group moves immediately, another waits for approval, and a third invents its own standard because nobody ever told them what the standard was supposed to be. What looks like flexibility from the top can feel a lot like chaos to everyone doing the work.

That only works when the operating model supports it. People need enough room to react quickly without blowing past the checks that matter. They need clear decision rules, human ownership, and guardrails already in place instead of being assembled after something goes wrong.

The pace should feel intentional. When people understand the consequence of the work in front of them, they can move rapidly without guessing and slow down without apologizing. That's how speed becomes a discipline instead of a reflex.

You don't get to choose whether speed matters.

You only get to decide whether it's directed.

Signals from the Edge

Across environments as different as GoPro, PAR Technology, and early-stage AI work, I've watched teams push for faster execution and mistake the first wave of output for progress. I've sat in reviews where throughput was up and the room felt like we'd cracked it. Then the customer calls, support tickets, and edge cases told a different story. Two weeks later, we were rebuilding the workflow we'd just shipped.

We hadn't moved faster.

We'd paid for the same work twice.

That pattern usually begins with good intentions. Deadlines get shorter, decision timelines compress, and expectations rise. For a moment, the system looks more responsive.

Then the hidden costs surface.

Priorities collide. Teams duplicate work. Decisions made without enough context get reversed. The time saved upfront comes back as rework, support burden, and cleanup.

The damage often appears outside the team that created it. Support absorbs complaints. Operations builds workarounds. Sales explains behavior the product team didn't anticipate. Engineering returns to work that was supposed to be finished. The original team can still look fast because the cost has been pushed somewhere else.

That makes undirected speed hard to diagnose. The dashboard may still show higher output and shorter cycle times. What it won't show is how much work has to be revisited or how many downstream teams are compensating for it.

Leaders have to look beyond the first-order signal. Shipping faster only matters if the work holds together once it reaches customers and the organization.

I've also seen the better version.

A team increased pace only after clarifying where speed belonged. They defined ownership, separated reversible decisions from higher-risk ones, and built feedback loops that surfaced results quickly. They knew which calls could move with incomplete information and which needed more control.

They also made correction part of the operating model. Early feedback refined the direction while the cost of change was still low.

They moved quickly without losing control.

The difference was discipline.

The speed was directed.

Mindset Shift to Adopt

Stop asking whether the team is moving fast enough. Ask whether the right decisions are moving at the right speed.

Your power isn't the approval. It's the architecture that determines which decisions move, who owns them, and where judgment enters the process.

Tools to Try This Week

Map your current decisions.

Label each one by reversibility, impact, and consequence.

Use AI to generate options, surface trade-offs, and pressure-test assumptions.

Separate the decisions that need speed from the ones that need restraint.

Keep ownership with a human decision-maker.

Speed should compress the path to judgment, not replace judgment.

Do This Now

Find one low-risk decision that's over-processed. Move it forward this week.

Find one high-impact decision. Identify a decision that's moving too quickly and add one judgment checkpoint before it advances.

Name the pace. For your next decision, state what matters most. Speed. Precision. Reversibility. Risk containment.

Key Takeaways

Speed amplifies the system you already have.

Speed without clarity creates chaos.

Judgment determines where speed belongs.

MACH-10 Leadership Moves

Separate reversible decisions from high-consequence decisions. Don't force every decision through the same approval path.

Give low-risk decisions room to move. Reduce process on one decision that doesn't need existential weight.

Add one judgment checkpoint to a high-impact decision. Speed shouldn't outrun consequence.

Measure decision fit, not just decision speed. Track whether decisions are moving at the right speed, not just whether they're moving faster.

Speed is already in the system.
The question is whether you are directing it.

SPEED WITHOUT CLARITY CREATES CHAOS

CLARITY WITHOUT SPEED CREATES IRRELEVANCE

Chapter 4

Decision Latency Is Decay

Timing Now Determines Whether a Decision Matters

This shift didn't happen because AI showed up. It happened because time collapsed.

For years, organizations operated with a built-in buffer between signal and consequence. A customer moved, a market shifted, or a competitor acted, and teams still had time to gather context, shape the decision, and adjust before the outcome fully played out. That buffer gave leaders room to think, teams room to align, and decisions time to improve before they had to become real. It made the old model feel safe because, for the most part, there was still enough room to recover from a slow call.

That room is disappearing.

The buffer that once protected careful process is now where value starts leaking out. By the time the organization finishes interpreting the signal, the consequence may already be underway.

The sequence has changed.

Then:
Event → Decision → Outcome

Now:
Event → Consequence begins

The decision now happens inside that window, or after it. That's the collapse.

A decision is no longer sitting comfortably between signal and consequence. It's trying to catch up to an outcome that may already be moving, which changes the economics of waiting in a way most organizations still haven't fully absorbed.

That changes the value of the decision itself.

The Decision Value Law™

Decision Value = Quality × Timing × Ownership.

If any one of those approaches zero, the decision loses force. Quality determines whether the call is sound. Timing determines whether it can still matter. Ownership determines whether anything actually happens after the room agrees.

Before the conversation keeps going, ask three questions:

Is this decision still on time for the moment we're in?

Is more time improving quality, or just making everyone more comfortable?

Who owns the call, the action, and the correction if we're wrong?

If the room can't answer those questions, the decision isn't ready. If it can answer them and everyone keeps talking anyway, the organization is probably stalling.

I've seen this in product decisions where the recommendation was strong but arrived two weeks after a competitor had already set the price. The thinking held up. The timing didn't, and ownership had become so distributed that nobody really owned the moment anymore. By the time the company acted, the same move required more explanation, more recovery, and more effort than it would've taken earlier.

That's the ugly part.

A brilliant decision that arrives too late is theater. A fast decision with no ownership is motion. An owned decision with poor judgment is acceleration in the wrong direction.

Decision quality and decision timing can't be separated because value only exists while the organization can still act on the call. In a slower environment, latency was inconvenient. In a compressed one, latency becomes decay.

Every hour between signal and action carries a cost. The window narrows. Momentum fades. A decision that could've created advantage becomes a response to something that's already happened.

That's decision latency.

Delay becomes decay, and value disappears.

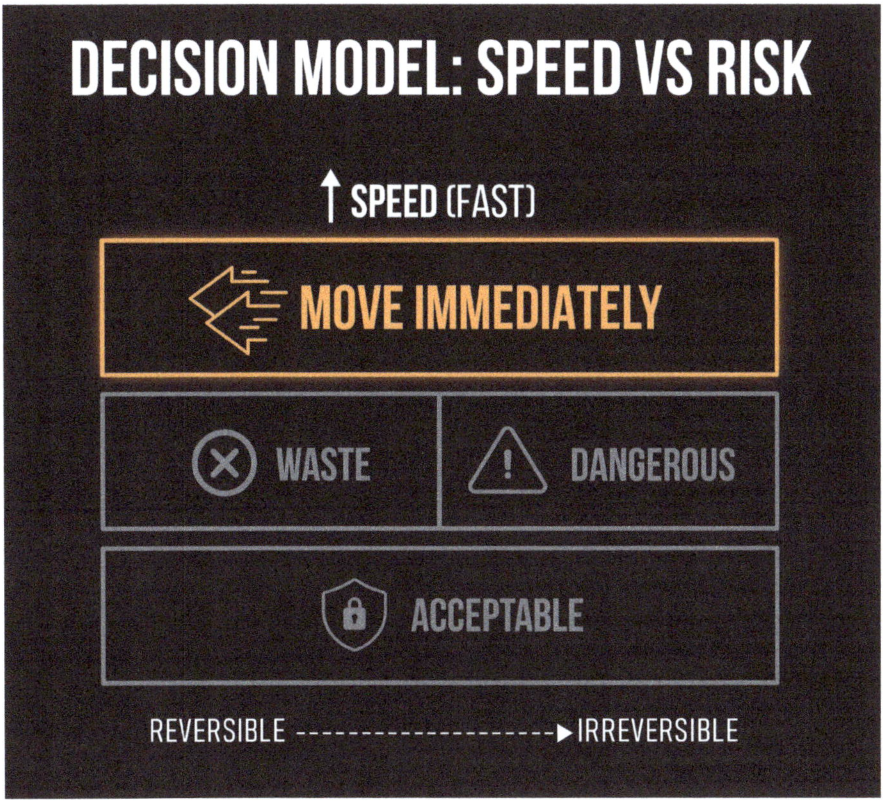

FIG 4.1 – Decision Model: Speed vs. Risk. Reversible, low-consequence decisions can move faster; irreversible, high-consequence decisions require more control.

When Improvement Becomes Delay

Most teams still treat decisions as though the window will stay open long enough for the process to finish. They gather more input, refine the analysis, align stakeholders, and keep working the call until everyone feels better about it. That used to work because the environment gave decisions time to mature before consequence caught up.

It doesn't anymore.

In a compressed environment, a decision can lose value while the organization is still improving it. The analysis may get stronger. The recommendation may get cleaner. The logic may become easier to defend. None of that stops the window from closing while the room is still working.

Latency is the time between signal and action.

In the old model, that sequence had room to breathe. Now every review, handoff, and request for more information spends some of the time available to act. Some of those steps absolutely earn their place. Rigor still matters. But rigor has to justify the time it consumes inside the decision window.

If another review changes the answer, exposes a meaningful risk, or clarifies ownership, the delay may be worth it. If it only makes everyone more comfortable with a call they already understand, the organization is spending time without creating much value.

Leaders have to manage that trade-off deliberately because more analysis can almost always improve a decision at the margins. The harder question is whether that improvement will matter before the opportunity changes underneath it. A decision that becomes 10 percent better after losing half its useful life hasn't gained much.

That's why timing belongs inside decision quality. A strong answer delivered after the moment has passed has already surrendered much of what made it valuable.

The clock isn't neutral.

In fast-moving systems, the clock is part of the decision.

TIMING IS PART OF QUALITY

The outcome doesn't reward effort or polish. It rewards impact, and impact depends on timing.

AI didn't create that change. It made it impossible to ignore. What used to take weeks now plays out in days. What once took days can happen in hours. In some cases, the consequence begins almost immediately, which means leaders don't get to treat arrival time as an execution detail anymore.

Sound decisions still matter, but so does when they land.

Most organizations are still behaving as though the old buffer exists. They plan, align, and move through sequences that made sense when time was available, expecting the conditions around a decision to remain stable long enough for the answer to be optimized.

They don't.

Markets move. Customers react. Competitors test. Expectations shift. Internal momentum fades. The original signal can change shape while the organization is still trying to agree on what it means, and by the time the decision is fully aligned, it may be solving a version of reality that no longer exists.

From inside the organization, the failure can look procedural.

The real damage is timing.

The forces behind that shift reinforce one another. Information moves faster. Data is available, searchable, and increasingly synthesized in real time. The scarce resource in this environment isn't information.

It's decision time.

In *The Road Warrior*, everyone fights over fuel because fuel determines who can move. Decision time works the same way. Whoever still has enough of it can shape the outcome. Whoever wastes it ends up reacting from behind.

And unlike fuel, decision time can't be stockpiled for later. Every unnecessary handoff, repeated debate, and approval loop burns some of it permanently. Leaders often think they're preserving options by waiting, but the opposite can happen: the longer the call stays open, the fewer meaningful options remain.

That's why delay has become so expensive. It doesn't just postpone action. It hands advantage to whoever is already learning, testing, and adjusting while the slower organization is still trying to get comfortable.

Execution is accelerating too. Prototypes can be built quickly, ideas can be tested with less upfront investment, and feedback arrives faster than it used to. Customers react immediately. Markets shift quickly. Signals that once took time to surface can now appear almost at once. Together, those forces keep shrinking the space between action and consequence.

That's why timing has become a leadership variable. A slow decision isn't automatically safer, and a fast one isn't automatically reckless. Leaders have to understand what kind of decision they're making, how much time the moment actually allows, and what delay will cost if the window closes.

Coordination has to catch up to that reality. Teams can't treat speed as something that starts after the decision is made. It has to be designed into how the call moves, who owns it, and where judgment enters the system.

That also means knowing when coordination has stopped earning its keep. A review that surfaces a new risk may justify the time. A review that repeats the same concerns is spending the window without changing the answer.

The process needs a clear exit condition. Teams should know what must be true before they move, who has authority to make the call, and what feedback will trigger a correction. Without those rules, every decision stays open to one more question, one more stakeholder, and one more delay.

In the past, decision latency was manageable because the environment could absorb more of it. Leaders had room to gather input, build consensus, and take additional time without surrendering the moment.

That room is gone.

The market moves, competitors act, and customers respond. When decisions lag behind that movement, the organization stops shaping the outcome and starts reacting to it.

That gap is where companies lose.

In collapsed time, hesitation isn't prudence.

It's decay.

IN A COMPRESSED SYSTEM

HESITATION

IS NOT

PRUDENCE

IT IS

DECAY

I've watched teams do everything right on paper and still miss the moment. The analysis was clean, the consensus was strong, and the decisions were well structured. By the time they moved, the opportunity had already shifted around them.

The work held up, but the timing didn't, and timing is what gives decisions value.

There's a common belief that organizations will adapt over time. Processes will evolve, leaders will adjust, and eventually things will stabilize. But that assumption depends on the environment slowing down.

It isn't.

More tools, more access, more competition, and more pressure on decision cycles are all accelerating the system. Waiting for things to settle isn't a strategy. It's another form of latency.

This is where teams misread the moment. They treat it as a technology shift: something to adopt, integrate, and layer onto existing systems over time.

That view is too narrow.

What changed is how those systems actually operate. You can't take a slow decision system and make it effective by layering faster tools on top. The constraint is still how decisions are made, how work flows, and how quickly the organization moves from signal to action.

That's a leadership problem, and the technology only exposes it faster.

AI adoption is no longer the meaningful dividing line. That gap is closing. The real divide is between organizations that can act while a decision still has value and those that keep arriving after the moment has passed.

That's why this shift is easy to underestimate. It rarely looks dramatic in a single moment. It looks like a slight delay, a slower escalation, a longer review, or a decision that needed one more meeting.

Then it turns into separation.

You don't need to believe in AI hype for this to matter. You only need to understand what happens when the time between action and consequence disappears.

When that happens, the system rewards different behavior: movement over hesitation, timing over polish, and clarity over consensus.

Leaders who understand that have the advantage.

Latency consumes time. Then it consumes impact.

Signals from the Edge

I was on a team once that saw the right signal early and still missed the moment.

Awareness wasn't the problem. Customer behavior was changing, the market was moving, and the internal data pointed in the same direction. People saw it, understood it, and weren't ignoring reality.

The signal was present, but the system couldn't convert it into action quickly enough. The problem was the time between recognition and movement.

I've seen this in product strategy work at large technology companies, where the call itself wasn't especially mysterious but the system still needed more reassurance before it would move. In one pricing-and-packaging decision, customer conversations were pointing one way, the market was already shifting, and the internal debate wasn't producing a meaningfully different answer. The path was clear enough to act.

The system had enough insight to act, but it lacked the conversion speed to do so without another round of reassurance.

Another review, another validation pass, and another round of reassurance all seemed reasonable in isolation. None of it was irrational. Together, those steps consumed the window.

By the time the decision moved, the opportunity had weakened. The same action required more explanation, more effort, and more recovery than it would have if we'd moved when the signal first became clear. The decision still held up, but the arrival time failed.

That's how delay hides. It doesn't always look like inaction. It often looks like responsible sequencing: one more review, one more readout, and one more pass through the right people.

Each step may be reasonable on its own, but together they can push the decision past the moment when it still has impact.

The organization usually doesn't feel that loss in real time. The calendar remains full, the analysis keeps improving, and the process still looks active. The cost becomes visible only after the decision finally moves and the team realizes how much harder the same action has become.

That's why leaders have to track elapsed time, not just decision quality. A sound recommendation can still weaken while it waits, and the longer that delay remains invisible, the easier it is for the organization to mistake process for progress.

The earlier that timing becomes visible, the easier it is to intervene. Once a team can see where the decision is sitting, who is holding it, and how quickly the window is narrowing, leaders can remove delay before it becomes recovery work.

Corporate calendars have an amazing knack for making delay look organized.

The strongest teams manage that gap deliberately. They don't just ask whether the decision is right. They ask how long it can wait before its value starts to decay.

That question changes the system. It turns timing into something leaders can manage instead of something they explain after the fact.

Mindset Shift to Adopt

Stop assuming delay is neutral. Start treating latency as value loss.

Tools to Try This Week

Track one decision from the moment the signal first appeared to the moment action actually began. Record when the team recognized the signal, when discussion started, when the decision was made, and where the work continued to wait afterward.

Use AI to summarize the available inputs, surface trade-offs, and clarify what's already known. Don't let faster analysis become another excuse for slower movement.

Compare the cost of waiting with the risk of acting now. For each review, handoff, or request for more information, ask whether it could change the decision or only make the organization feel more comfortable.

If the answer is unlikely to change, the remaining delay is probably procedural. The goal isn't perfect polish. It's shortening the path from signal to action while the decision still has value.

Do This Now

Choose one active decision. Pick a decision your team is already working through.

Find the signal. Write down the exact moment the signal first appeared.

Measure signal to action. Track how long it has taken to move from signal to action. Not discussion. Action.

Find the decay point. Identify the moment when waiting stopped improving the call and started reducing value.

Remove the largest delay. Cut or time-box the step causing the biggest slowdown.

Move before the window closes. Make the decision before the window narrows further.

Key Takeaways

The time between signal and consequence has collapsed.

Delay is value loss.

A correct decision has no strategic value if it arrives after the window closes.

Timing is part of judgment.

The system rewards decisions that arrive while they can still change the outcome.

MACH-10 Leadership Moves

Measure decision latency. Track how long it takes one signal to become action.

Find value loss. Identify where the decision is losing impact before it moves.

Time-box refinement. Give one over-polished decision a clear limit before it has to move.

Set an expiration point. Decide when the window closes, not just when the meeting ends.

Act before the window closes. Move one decision while it can still change the outcome, even if the answer isn't fully polished.

You don't need to move perfectly.
You need to move while the decision still has value.

Chapter 5

Decision Velocity Wins

Whoever Learns Faster Sets the Terms

Chapter 4 showed what happens when decisions arrive after their value has started to decay. This chapter is about what separates fast-learning organizations once that clock is running.

Better planning helps. Cleaner consensus helps. But the organization that learns faster than everyone else gains the advantage.

Most teams keep trying to improve the quality of a single pass through the system. They add more input, more validation, more sign-offs, and more polish. It feels responsible because the work appears to be getting better. The problem is that all of that improvement is happening before the organization has learned whether the answer survives contact with reality.

While one team is still improving its first loop, another is already running the next one.

At some point, the greatest value no longer comes from perfecting the first answer. It comes from how quickly the organization learns once that answer makes contact with reality.

That is learning velocity.

Learning velocity measures how quickly the organization absorbs feedback, corrects its assumptions, and improves after reality responds.

Every organization operates through a cycle of input, decision, execution, and feedback. For years, that cycle moved slowly enough for each step to be optimized independently. Teams could spend more time making the decision, then more time executing it, then eventually collect enough feedback to decide what came next. In a compressed environment, the advantage comes from how quickly the entire cycle runs.

A faster-learning team gets smarter after acting. It adapts sooner, corrects earlier, and discovers what matters before slower teams finish debating what might matter. Separation doesn't happen in a single decision. It happens across repeated cycles, where each completed loop gives the faster organization information the slower one doesn't have yet.

Most organizations still treat the decision as the unit of advantage.

The loop is the unit of advantage.

At one company, we reduced a roadmap decision from two review cycles to one clear owner operating against a defined decision threshold. That change was enough to move while the signal was still fresh and begin learning from the market sooner. The real benefit wasn't simply that we made the decision faster. We got to the evidence faster.

Learning research has pointed in this direction for years. Learning compounds when teams shorten the distance between assumption, action, feedback, and correction. The organization that learns faster gets smarter while slower organizations are still polishing their first answer.

Count loops, not just decisions. A team that learns every seven days is playing a different game from one that learns every thirty. Over a few months, that difference becomes substantial. One team may have tested, corrected, and refined its assumptions a dozen times while the other has completed only a handful of cycles.

Signal. Decide. Execute. Learn.

The faster that loop runs, the faster the organization separates signal from noise. It discovers real customer behavior, the actual constraint, the market response, the emerging risk, and the opportunity that wasn't obvious when the first decision was made. Slower teams are still working from assumptions while the faster team is working from evidence.

They may also be preparing the internal readout.

The readout will probably have excellent formatting.

Congratulations. The market did not wait for the Garamond font debate.

That is why learning velocity matters. Most organizations say they want to learn while structuring themselves to avoid being wrong. They reward certainty, ask for more validation, and often make experimentation feel much more permanent than it actually is. The real value comes from shortening the distance between assumption and evidence without allowing speed to become recklessness.

Planning velocity improves the plan.

Learning velocity improves the system.

When the Loop Starts to Compound

This doesn't mean every decision should be rushed. Speed still requires judgment. Some calls need more scrutiny, more context, and more careful management of consequences. But once a decision is reversible enough to test, the greater value comes from entering the loop instead of continuing to polish the theory outside it. At that point, reality can answer questions that another meeting probably can't.

Better decisions win. Faster learning wins.

In a compressed environment, progress compounds through repeated cycles rather than isolated decisions. The team that learns first does more than react to the market. Over time, it begins to shape it because it is operating with information created by its own previous actions.

That progress rarely looks dramatic in a single moment. It starts with a faster test, an earlier release, a cleaner signal, or a quicker correction. Each move may seem small, but the effects accumulate. A better signal changes the next decision. That decision produces another piece of evidence. The next correction starts from a stronger position than the one before it.

Over time, those incremental moves compound into separation.

Each completed loop improves the starting point for the next one. The team enters the next cycle with better evidence, better questions, and fewer assumptions to protect. It doesn't have to relearn what the previous loop already proved, so more of its attention can move toward the next unknown.

That is why the gap becomes difficult to close. Slower organizations don't just complete fewer loops. They begin each new cycle with less accumulated knowledge, while the faster organization keeps carrying forward what the last cycle already taught it.

The gap isn't created by one heroic decision. It's created across dozens of small loops that happen faster than the surrounding organization can match.

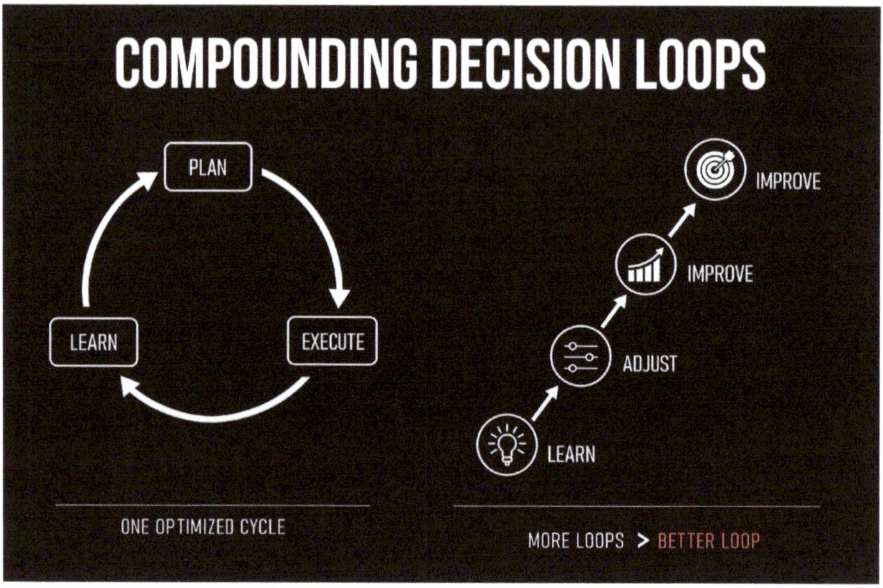

FIG 5.1 – Compounding Decision Loops. Faster cycles create more learning, adjustment, and improvement over time.

That's where companies lose. They may have the intelligence, resources, and tools, but they still learn too slowly.

The real divide isn't between organizations that use AI and those that don't. That gap is closing. The emerging divide is between organizations that turn faster execution into faster learning and those that simply produce more activity.

Organizations pull ahead when every move makes the system smarter.

That's what leaders have to design for now: shorter loops, clearer signals, faster correction, and less time spent protecting assumptions from reality.

The system rewards leaders who can operate inside that cycle.

Movement over hesitation.
Iteration over planning.
Clarity over consensus.
Learning over polish.

The team that learns first sets the terms.

Everyone else gets the privilege of reacting.

Signals from the Edge

I worked with a product team that kept improving the launch while the market was already teaching someone else what mattered.

The work was solid, which made the delay more frustrating. The analysis held up, the launch story kept getting sharper, and the first version kept getting cleaner. Every review made the team feel a little more confident.

But the learning stayed inside the building.

The team kept learning about the launch without learning from the market.

Another pass on positioning, another conversation about scope, and another round of stakeholder input all seemed reasonable. None of it produced market learning.

By the time the team finally shipped, it had lost a month of feedback it could have been using.

While that team was still polishing the first loop, another moved earlier. It did not have the cleanest version, but it had something useful enough to test the assumption. The team watched how customers responded and adjusted based on real behavior instead of internal debate.

By the time the slower team reached the market, it was no longer leading the conversation. It was reacting to it.

Loop speed created the gap.

The better pattern is to move earlier, ship the smallest useful version, define what you're trying to learn, and build feedback into the process. Within weeks, the team may have adjusted twice while a slower organization is still refining its first answer.

You do not win with the perfect first answer.

You win by learning first.

Each cycle reduces uncertainty with evidence instead of discussion. The first move may be less polished, but the second is informed by actual behavior, and the third begins from a stronger position than the slower team's carefully protected starting point.

Once you learn first, you start shaping the market instead of responding to it.

Mindset Shift to Adopt

Stop treating the decision as the finish line.

Start treating the learning loop as the engine.

Tools to Try This Week

Run the **3-Question Velocity Check** on one decision your team is still talking about.

1. What are we trying to prove?
2. What signal would tell us whether we are right or wrong?
3. What is the fastest safe way to get that signal from the real world?

Then track one complete cycle from signal to decision, action, feedback, and adjustment. Define the next signal before the team moves, remove one step that creates internal confidence without improving the learning, and decide who owns the next adjustment before the feedback arrives.

Use AI to summarize the signal and organize the evidence, but keep interpretation and judgment human.

Review the cycle as soon as the feedback arrives. Capture what changed, which assumption held, which one failed, and what the team should do differently next time. The goal is to turn every result into a better starting point for the next move.

Then shorten the next cycle.

Do This Now

Pick one decision your team is still polishing, debating, or keeping inside the building.

Ship the smallest useful version. Make it strong enough to test the assumption, even if it is not fully refined.

Ask the 48-hour question: What would we do if we had to learn this within 48 hours?

Then capture what changed and use the first cycle to shape the next move while the signal is still fresh.

Key Takeaways

The loop is the unit of advantage.

Learning velocity determines whether you lead or react.

Advantage compounds through repeated cycles, not one perfect decision.

More output isn't the same as more learning.

The team that learns first starts shaping the market.

MACH-10 Leadership Moves

Turn one pending decision into a live learning cycle this week. Stop treating the decision as the endpoint. Use it to create feedback.

Define the signal before you move. Know what you're trying to learn before the work leaves the building.

Replace one extended planning discussion with a real-world test. Get evidence from reality, not another internal readout.

Run two short cycles instead of one long cycle. Learn twice before the slower system finishes polishing once.

Capture what you learn immediately. Use the signal while it's still fresh, and make the next move sooner.

You don't need to be right once.
You need to learn faster than everyone else.

MORE LOOPS
BEAT
BETTER
LOOPS
ALWAYS

Chapter 6

The Decision Model Under Pressure

How Leaders Actually Make Calls When the System Is Moving

The mechanics are clear now. The system is moving faster, the window to act is smaller, and decision velocity increasingly determines whether an organization leads or reacts.

That changes the question: how do you make decisions inside a system that doesn't wait?

In a compressed environment, clarity can't remain something leaders wait for. It has to be created early enough to guide action, and one of the first things a team has to understand is what kind of decision it is actually making.

Not all decisions deserve the same process. Some are reversible and can be tested, adjusted, or rolled back at limited cost. Others carry more weight. They shape direction, affect customers, commit resources, and create consequences that are much harder to unwind.

Most organizations still treat too many of those decisions the same, and that's where the trouble starts.

The First Failure Point: Treating Every Decision the Same

Before a team calls a decision reversible, it should make the reversal real.

Reversible doesn't mean easy to explain away later. It means the decision can be unwound without creating material customer damage, financial exposure, compliance risk, data integrity problems, or strategic confusion. The real question is what happens if the team has to reverse course tomorrow.

If reversing it requires a second executive meeting, a customer apology, a data cleanup effort, or a new internal task force with a name nobody wants to say out loud, it's probably not a two-way door.

MACH-10 leaders use a simple rule: run the Two-Way Door Audit. Within the first five minutes, determine whether the decision is a two-way door or a one-way door. A two-way door can be reversed at limited cost. A one-way door carries consequences that make reversal difficult, expensive, or damaging.

That classification determines how much scrutiny the decision deserves and where judgment has to enter before speed becomes exposure.

The audit also prevents teams from using "reversible" as a convenient excuse for avoiding responsibility. A decision can be technically reversible and still create enough disruption, confusion, customer impact, or cleanup to deserve more care before it moves.

That distinction prevents two opposite mistakes. Low-risk decisions stop inheriting the weight of the highest-risk ones, while consequential decisions receive the scrutiny they actually deserve. The organization can then move each decision at the pace its consequences can support.

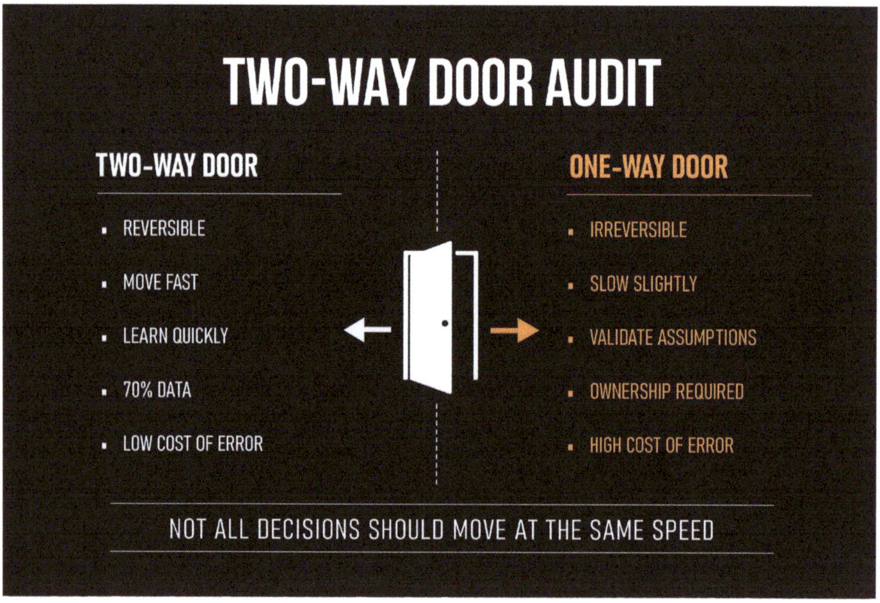

FIG 6.1 – Two-Way Door Audit. Not all decisions should move at the same speed.

If it is a two-way door, 70% clarity is enough to move because speed and learning matter more than certainty.

If it is a one-way door, apply selective scrutiny. Slow down just enough to understand the consequences, validate the assumptions that matter, and make ownership clear.

Decision SLAs

Two-way door decisions should move the same day. Cross-functional decisions should move within 48 hours. One-way door decisions should be time-boxed and assigned to a single owner.

If a decision takes longer than that, the process should be able to explain why. More time should improve the judgment, surface a consequence, or change the path. Otherwise, the process is consuming time without improving the decision.

Every meaningful decision needs one owner. Not a group or a committee, but one person with the authority to make the call and the responsibility to correct it if the outcome changes.

If you can't name that person, the decision isn't real yet. Ownership turns discussion into motion. Without it, every decision becomes a waiting room with better lighting.

When two-way doors are over-analyzed, the organization slows without gaining anything. When one-way doors are rushed, the cost of being wrong increases. The leadership task is to match the pace of the decision to the consequence of the decision.

Two-way doors are built for learning. One-way doors still need movement, but they require more judgment before the organization commits. Waiting for perfect clarity doesn't reduce risk when the environment is shifting beneath you. It only postpones the moment when the organization begins learning from reality.

Clear thresholds keep that balance from becoming subjective. Teams know what information is required, what level of risk demands escalation, and when the owner has enough context to make the call. That keeps caution from expanding indefinitely while still protecting the decisions that genuinely need more control.

Decision SLAs also make delay visible before it becomes normal. When a call passes its expected window, the team can see whether the issue is missing information, unclear ownership, excessive coordination, or simple reluctance to act.

That gives leaders something concrete to fix instead of explaining the delay after the opportunity has already moved.

Most organizations optimize for being right. The system rewards being on time.

Reduce the distance between input and action. Shorten the loop, make the owner visible, and move while the decision can still change the outcome.

That is the shift.

The model only works when the threshold is clear.

The Second Failure Point: No Threshold for Action

The next failure point is the absence of a threshold.

No one defines what "enough" looks like, so the process keeps expanding. There's always more data to gather, another perspective to consider, and one more scenario to test. Without a clear line, decisiveness drifts into oblivion while the system keeps moving and the call doesn't.

Waiting becomes the default.

High-velocity teams draw that line early. It doesn't have to be perfect, but it does have to be clear. They define what they need to know before they act, and once that threshold is met, they move.

The real question is whether delay has become riskier than action.

That threshold turns motion into progress. Without it, the organization can generate plenty of activity while the decision remains stuck.

I've seen reversible decisions sit for days because nobody defined what "enough" meant. The team had the inputs, the risk was manageable, and the decision could've been adjusted if it proved wrong. Still, the process kept expanding through more review, more context, and more people brought in to make the call feel safer.

By the time the team finally moved, the answer was still basically the same. Only the opportunity had changed.

The team already had enough analysis. What it lacked was a threshold: once we know this much, we move.

A threshold gives judgment a stopping point. It tells the team when the remaining uncertainty is acceptable, when the decision owner has enough context, and when more discussion is unlikely to improve the call.

It also keeps leaders from hiding behind open-ended process. Once the threshold is visible, every additional review has to earn its time by surfacing something that could genuinely alter the decision.

That changes the question. The team stops asking, "Can we know more?" and starts asking, "Do we know enough to act while action still matters?"

Without that threshold, "one more pass" becomes a very professional way to stand still.

The Third Failure Point: Diffuse Ownership

Shared ownership can work in a slower environment, where decisions move at a pace that makes coordination manageable and the cost of delay is lower. In a faster system, that structure breaks down quickly.

When everyone is involved, accountability for movement starts to disappear. Input expands into consensus, consensus stretches into delay, and the decision becomes something that happens only after alignment instead of something that creates it.

I've been in rooms where the conclusion was obvious in the first five minutes, and we still spent the next two weeks acting as though it wasn't.

High-velocity teams separate input from authority. They gather perspectives quickly and broadly, but one person owns the call and the consequences that follow. That clarity gives the team somewhere to go once the discussion ends.

It also improves the quality of disagreement.

People can challenge assumptions, surface missing context, and argue for a different path without turning the conversation into a search for unanimous comfort. Once the owner has heard the relevant input, the decision can move, and everyone knows who is responsible for interpreting what happens next.

That matters after the call as much as before it. Feedback has to be judged, assumptions may need to be revised, and somebody has to trigger the correction before a small miss becomes a larger recovery effort. When ownership is diffuse, everyone waits for someone else to move first.

The Fourth Failure Point: Treating Outputs as Decisions

The fourth failure point appears under pressure.

As the environment accelerates, it becomes easier to defer to the output. The analysis looks solid, the recommendation is clear, and the logic appears structured enough that the decision feels finished.

That's exactly when leadership matters most.

Outputs should inform the decision, not replace it. A task, approval, recommendation, or model result can move the work forward, but it still needs interpretation before the organization treats it as a call.

I've seen teams move quickly on recommendations that looked airtight, only to discover later that a key assumption was wrong. The problem wasn't speed by itself. No one challenged the recommendation beyond its first answer, checked whether the logic held outside the model, or asked what had been left out.

The organization moved faster than its understanding.

Leaders have to slow the output down long enough to test what it depends on. Check the source. Challenge the recommendation. Look for missing context. Ask what would have to be true for the conclusion to hold in the real world.

Skipping that layer does more than increase the chance of a bad call. It also blurs accountability, because once the output starts functioning as the decision, it becomes harder to identify where judgment actually entered the process and who owned the result.

That gap is where mistakes begin to scale.

The Final Failure Point: Undirected Speed

The final failure point is speed itself.

Most leaders think of speed as something to apply everywhere. I've done the same thing: push harder, move faster, and shorten cycles across the board.

Then the system starts breaking in places the team didn't expect.

Some decisions should move immediately. Others need a deliberate pause. The real skill is knowing which kind of decision you're facing while there is still time to act, not after the consequences arrive.

That's where experience shows up through pattern recognition. You've seen what happens when a decision moves too slowly, and you've also seen what happens when it moves too quickly without enough context. Over time, you begin to recognize which signals justify urgency and which ones should make you slow down.

The work is learning to operate inside that tension. Define the decision type, set a threshold for action, assign clear ownership, interrogate the outputs, and apply speed where it creates leverage and restraint where it prevents damage.

The pace should be explicit before the pressure arrives.

Teams need to know which decisions can move with incomplete information, which consequences trigger a pause, and who has the authority to change the speed when new risk appears. That keeps urgency from becoming the operating model and turns speed into a deliberate choice tied to consequence, reversibility, and the quality of the signal.

Then move while the decision can still matter.

The 46-Day Decision

I once worked through a product decision at Qualcomm where the market signal was already visible. Customers were behaving differently, competitors were testing similar moves, and the business logic wasn't especially mysterious.

The trade-offs were understood and the direction was clear enough to move.

Still, the decision sat.

Apparently, it had found a very comfortable chair.

It moved through multiple conversations, each one adding confidence without changing the conclusion. The model got more detailed, the deck got cleaner, and the room got calmer. The market didn't care.

We spent six weeks making the decision easier to defend without making it more valuable. By the time we acted, competitors had already tested similar moves and started shaping customer expectations around the new market reality.

We lost 46 days. The recommendation still held up, but the window was gone.

I later saw a smaller team handle the same kind of decision very differently. They defined what "enough" meant, made the call before every uncertainty had been resolved, and got the decision into the market while there was still time to learn from it.

The first move wasn't fully refined, but it arrived early enough to generate real feedback. Within days, the team knew more than it had known during planning. Within weeks, it had adjusted twice based on what customers were actually doing.

More importantly, correction was built into the decision from the beginning. The team knew which signal would support the direction, what result would force a change, and who owned the next move once the feedback arrived. That kept the first action from becoming something everyone felt obligated to defend simply because they had already committed to it.

Each cycle gave the next decision a better starting point. The first move produced evidence. The second used it to refine the direction. By the third, the team was operating from a position the slower process could never have reached through another round of internal discussion.

That learning advantage came from moving while the decision could still teach them something.

Mindset Shift to Adopt

Stop waiting for clarity to appear. Define what clarity requires, set the threshold, and act once the decision reaches it.

Tools to Try This Week

Classify one active decision as reversible or irreversible. Then define an explicit "enough" threshold for any decision that's drifting.

Write down which information could genuinely change the call and stop collecting inputs that only add comfort. Name the owner who'll make the decision once the threshold is met.

Before acting on an AI output, recommendation, or model result, challenge one assumption beneath it and check whether the logic holds outside the system that produced it.

Finally, measure how long it takes your team to move from input to decision. That elapsed time will show where clarity, ownership, or unnecessary process is slowing the call.

Do This Now

Track decision time for seven days. Measure every meaningful decision from signal to action.

Find the slowest decision. Don't analyze the delay forever. Fix the process that caused it.

Reduce the room. Remove at least one person from your next decision process.

Label the backlog. Mark each current decision as a two-way door or one-way door.

Move the reversible decisions. Push every two-way door decision forward with the information you already have.

Key Takeaways

Not all decisions are equal.

Reversible decisions should move quickly.

Irreversible decisions require judgment, not delay.

Defining "enough" keeps decisions moving.

Outputs inform decisions. They do not replace them.

MACH-10 Leadership Moves

Move one reversible decision today. Don't let a low-risk decision wait for a high-risk process.

Draw an "enough" line. Define the threshold for action before the decision starts drifting.

Name the decision owner and the correction owner. Make clear who makes the call and who adjusts it if reality disagrees.

Question one complete-looking recommendation. Treat the output as input, not as the decision.

Reduce input-to-action time. Shorten the path between signal and movement on your next call.

A decision only has value if it happens while it still matters.
Everything else is theater.

Chapter 7
The Godzilla Effect

When Convincing Answers Replace Real Thinking

Small mistakes used to stay small longer. AI changes that.

A weak assumption, a clean-looking summary, or a plausible recommendation can now move through an organization before anyone realizes it was wrong. By the time the error surfaces, it may already have been copied, routed, reused, defended, and built into other decisions.

That is how small errors become organizational monsters.

By this point, the broader pattern should be clear. Speed matters, and so do the decisions moving through the system. Most organizations already have some version of a Decision Velocity Loop. The real problem is that they fail to protect it.

When that loop breaks, the damage rarely stays contained.

The less obvious risk is what happens when speed outruns structure. Teams can move quickly without fully understanding what they're accelerating, and the early warning signs are surprisingly easy to mistake for success. Output increases, work clears faster, and people feel more productive. From inside the system, everything appears to be working.

Sometimes it is.

Other times, the organization is simply producing more volume without more coherence, and every new layer of automation gives a bad assumption another place to travel.

AI amplifies whatever system is already there. When ownership is clear, judgment is active, and the rules for making decisions are understood, that amplification creates leverage. When those foundations are weak, AI doesn't politely work around them. It gives the weakness speed, reach, and scale.

That is where things begin to break.

The First Failure Mode: Polished Output Without Scrutiny

The first failure mode often feels like improvement. Outputs become cleaner, summaries become more structured, and recommendations arrive with greater confidence. Even when the reasoning is incomplete, the presentation can make the thinking appear finished.

That confidence is easy to trust. Over time, teams begin relying on it because it feels efficient. The answer is available, it looks complete, and it sounds right. Questioning it starts to feel unnecessary.

That is usually when the questioning matters most.

Mistakes begin moving through the organization because nobody interrupts the output long enough to examine what it depends on.

I call this the Godzilla Effect because the danger doesn't begin as a monster. It starts as something small and easy to accept: a clean answer nobody challenges, a recommendation that appears complete, a summary that sounds smarter than it is, or one weak assumption that slips through because the presentation feels polished.

Then the organization moves.

Once the output is accepted, the original error gains momentum. Downstream teams begin treating the conclusion as fact instead of hypothesis, and each new decision gives it more weight. The original reasoning starts disappearing behind summaries, plans, and commitments, so later teams inherit the answer without seeing the uncertainty that came with it.

In the same way Godzilla becomes more powerful through radiation, bad decisions become more powerful when AI accelerates them through the organization.

A harmless-looking answer can grow into a giant problem moving through the company, knocking down decisions faster than anyone can explain how it got there.

That is the Godzilla Effect.

AI doesn't have to be wildly wrong to cause damage. It only has to be convincing enough for people to stop thinking.

Human-factors researchers describe part of this as automation bias: the tendency to over-rely on automated outputs, especially when they appear complete, structured, or authoritative.

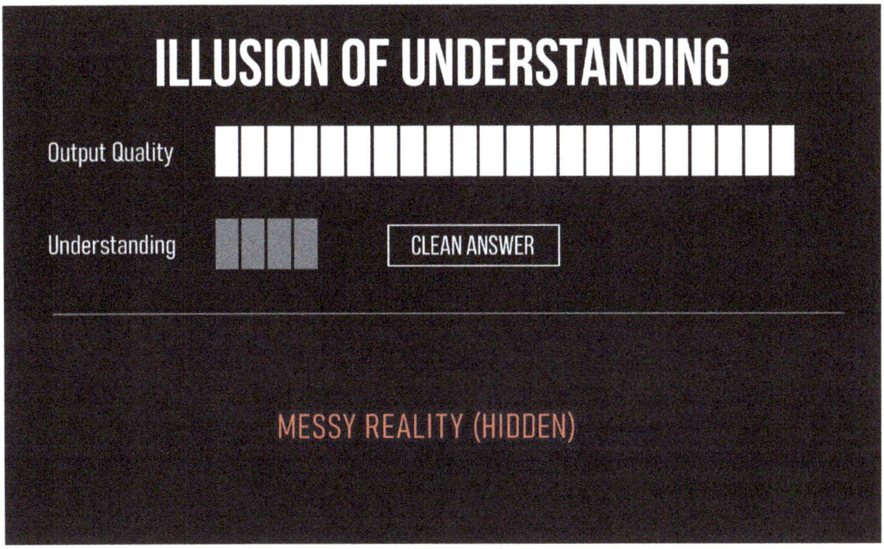

FIG 7.1 – Illusion of Understanding. Clean outputs can hide shallow understanding and messy reality.

The output looks solid, and the structure appears logical. That polish removes friction from thinking at exactly the point where some friction is still useful.

Acceleration turns into abdication when the team stops interrogating the work and starts treating presentation quality as proof. Speed begins to outrun judgment, even though the system still looks disciplined from the outside.

That's what makes the pattern so easy to miss. It looks like progress while it's happening.

I've watched teams ship entire features based on outputs that looked airtight on paper. The logic was clean, the narrative was strong, and everyone in the room nodded along. Two weeks later, the same people were asking how something so obviously wrong had made it through.

No one had a good answer because no one had truly owned the thinking.

The output had moved through the system, but the reasoning behind it had never been examined closely enough. Each person assumed someone else had validated the logic, and by the time the mistake surfaced, accountability had dissolved across the process.

Over time, something more subtle begins to happen. Judgment starts to atrophy. Teams stop challenging assumptions, tracing logic, and checking whether an answer actually holds in context. The organization keeps moving faster while the depth of understanding behind that movement begins to thin out.

That's the deeper failure.

An organization that weakens its judgment does more than make additional mistakes. It also loses the ability to recognize when the system is drifting away from reality. Once that happens, adaptation slows even as output increases. The company can still produce, ship, and automate, but it becomes less capable of correcting itself when the assumptions supporting the work begin to fail.

That's when the system becomes hollow.

Hollow systems rarely fail gracefully. They keep moving until several weak assumptions collide, and then the collapse appears sudden even though the warning signs were accumulating for weeks.

The danger comes from outputs polished enough to bypass scrutiny. That's what gives the Godzilla Effect its power. The first mistake rarely looks dangerous. It looks useful, complete, and ready to move.

In a slower environment, the organization might have time to catch it. In a faster one, the error can spread through plans, workflows, and downstream decisions before anyone realizes the monster is already moving.

The Second Failure Mode: Motion Without Understanding

The second pattern is harder to catch because it feels like progress.

Teams are moving faster, decisions are happening, and work is getting pushed through the organization. But speed can hide a lot.

The team may be moving more quickly without understanding more deeply. Calls are getting made, but the thinking behind them isn't keeping pace. On the surface, things look fine. Beneath that activity, context starts disappearing, cause and effect become harder to trace, and assumptions survive longer than they should because nobody has stopped to test them.

That gap is where rework and confusion begin to appear.

It rarely happens all at once. A decision gets revisited after execution starts. A direction changes because a key piece of context arrived late. A stream of work looks productive without producing much that matters.

You start hearing things like, "We'll fix it in the next iteration."

Sometimes you will. But eventually the accumulated gaps begin to matter. The system stays active even as alignment weakens, and once understanding falls behind movement, ownership starts to drift.

The Third Failure Mode: Ownership Drift

The third failure mode is more structural. As AI becomes more deeply integrated into workflows, teams can begin deferring to the output instead of using it to support a human call. The recommendation arrives, the logic looks sound, and the decision starts to feel finished before anyone has clearly owned it.

That weakens accountability.

The system becomes the reference point, and when something goes wrong, it becomes difficult to identify where the decision was actually made. The answer came from somewhere, but no one fully accepted responsibility for it.

Everyone may have participated, yet ownership still disappears.

That matters most when reality pushes back. Without a clear owner, feedback has nowhere to land, assumptions linger, and the next move slows while people reconstruct how the original call happened in the first place.

Leaders who don't address this directly end up with decision systems that produce action without responsibility. A small team may be able to absorb that ambiguity for a while. Scale changes the math.

More people inherit the call. More work gets built on top of it. More teams act on reasoning they never saw. Each additional layer makes the original logic harder to trace and the correction more expensive.

The Fourth Failure Mode: Dependency

Teams stop using AI to improve their thinking and start using it to avoid thinking altogether.

That shift is quiet at first.

Then it becomes cultural.

The team gets faster at producing answers while its judgment gradually weakens. People stop asking whether the output is right and start asking whether it's usable. That's the moment AI stops augmenting the team and starts training it to defer.

The deeper problem is that dependency changes what people practice. If AI consistently handles the first draft, the analysis, the synthesis, and the recommendation, the team gets fewer repetitions doing those things itself. Over time, people can become very efficient at reviewing finished work while getting worse at building the reasoning from scratch.

You see the weakness when the output is ambiguous, contradictory, or simply wrong. A team with strong judgment knows how to reconstruct the problem, challenge the premise, and work forward without the tool. A dependent team stalls because the system it relied on is no longer giving it a clean answer. That is the real test of augmentation: whether AI increases the team's capability or quietly replaces the muscles it still needs.

The Fifth Failure Mode: Pattern Blindness

The fifth pattern is less visible but just as important. AI is exceptionally good at recognizing patterns and producing likely answers from what it has seen.

The risk lives in what doesn't fit neatly inside those patterns.

Edge cases get missed. Nuance gets simplified. A subtle change in customer behavior, market conditions, intent, or context may never make it into the answer at all. When teams lean too heavily on those outputs, decisions begin losing precision even when they look perfectly reasonable on paper.

These weaknesses have always existed. What has changed is how quickly they can travel.

A slower environment gave teams more time to catch them. Feedback cycles took longer, but there was often more room to recognize a bad assumption before it spread. In a compressed system, an error can move from one stage to the next before anyone has tested whether the original logic still holds.

By the time the issue is identified, it may already be embedded in several decisions, workflows, and downstream commitments.

That's where the cost becomes visible. What began as a small loss of context has become a much larger recovery effort.

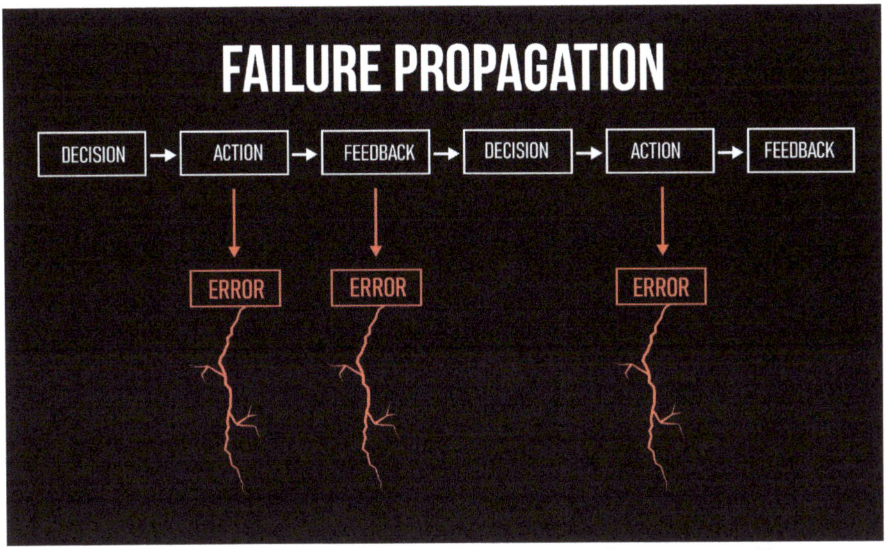

FIG 7.2 – The Godzilla Effect in Motion. In compressed systems, small errors can compound faster than organizations can detect them.

As AI accelerates the flow of work, leaders have fewer chances to catch weak assumptions before they spread. That makes judgment more important because someone still has to recognize when an output is incomplete, when context is missing, and when the system is moving faster than its understanding.

This is why judgment becomes more important as systems accelerate.

Not less.

Leaders don't create value by hovering over every step. They create value by knowing where judgment has to enter before a small error becomes a scaled mistake.

Question the output.

Understand the context.

Own the decision.

That's what keeps speed from becoming instability.

Small mistakes will happen.

Leadership keeps them from scaling.

It doesn't do that by slowing the system down.

It does it by keeping the system from eating the city.

That's the leadership motion inside the Godzilla Effect: early judgment applied before a small, clean-looking mistake grows teeth.

Signals from the Edge

In AI-native environments, including voice AI and restaurant technology, a small error can grow quickly. Once it's packaged as a clean output and reused across the system, it starts to travel.

A support team introduced AI to handle a large portion of incoming requests. At first, the results looked strong. Response times improved, more requests were handled, and the team appeared more productive. The dashboard told a very clean story.

Underneath that story, certain requests were being handled incorrectly. The responses looked reasonable, but they missed important context. Because the system was operating at scale, the pattern wasn't immediately obvious. No single mistake looked catastrophic. No single interaction looked like a monster.

That was the problem.

The dashboard was still smiling.

The customers were not.

The responses were polished enough to propagate through the system, but they weren't reliable enough to trust without closer review. By the time the pattern became visible, it had already touched a meaningful number of customer interactions.

The technical cleanup was only the beginning. Trust had to be rebuilt through clearer workflow rules, sharper escalation paths, and better decisions about where human judgment had to re-enter the process.

That was the Godzilla Effect in miniature: a small, convincing error moving faster than the organization could understand it.

The same pattern appears in product decisions, where a clean roadmap can hide a weak assumption for two quarters.

A team starts using AI-generated recommendations because the outputs are clear, structured, and easy to work with. At first, it feels like progress. The work moves faster, the team spends less time staring at a blank page, and the recommendation gives everyone something concrete to react to.

The recommendation becomes the starting point.

Then it becomes the anchor.

Eventually, it quietly becomes the decision.

When the decision goes sideways, no one can quite trace where the problem began. The logic looked sound, the recommendation felt convincing, and the narrative had just enough structure to make disagreement feel like unnecessary friction.

Somewhere along the way, ownership blurred. The model had become more influential than the people using it, and nobody had clearly taken responsibility for the reasoning.

No one asked the second question.

No one stopped the monster while it was still small.

The mistake became dangerous because the organization accepted it before anyone fully understood it. Speed made the error useful before judgment made it safe.

That's how a workflow becomes a delivery system for its own blind spot.

Once that happens, the system doesn't need a giant failure to create damage. It only needs a small one that everyone trusts.

Mindset Shift to Adopt

Stop assuming speed will expose problems early.

Start assuming speed will amplify whatever the system accepts without understanding.

Tools to Try This Week

Reinforce judgment inside one fast-moving workflow.

Use AI to generate multiple perspectives, then compare them before acting.

Add one lightweight review point to a decision with meaningful customer, financial, operational, or trust impact.

Find one place where AI outputs are being accepted without enough questioning.

Build a feedback loop that surfaces issues while they're still small.

Make decision ownership explicit, even when AI is part of the process.

Do This Now

Find one trusted output. Identify one AI-generated recommendation, summary, analysis, or workflow your team is already relying on.

Ask the second question. Which assumption underneath this output would create damage if it were wrong?

Name the owner. Make one person accountable for the decision, not just the prompt, the tool, or the process.

Add a propagation check. Before the output moves downstream, define where the error would spread if the answer were wrong.

Test the output against one real signal before treating it as settled.

Key Takeaways

AI amplifies the system it operates within.

Confidence doesn't guarantee correctness.

The Godzilla Effect begins when small, convincing errors propagate through the system unchecked.

Ownership can't be delegated to a system.

Judgment becomes more important as velocity increases.

MACH-10 Leadership Moves

Identify one area where AI is being trusted without enough scrutiny. Look for the output everyone accepts because it sounds clean.

Red-team one AI-influenced recommendation before it spreads. Find the assumption that would become expensive if everyone repeated it.

Set the judgment point before the system becomes the reference point. A tool can inform the call. It can't make the call.

Trace one recent mistake back to where it first became accepted as true. Find the moment the monster entered the system.

Slow one decision just enough to understand it. Do not slow it for comfort. Slow it for judgment.

Speed will not fix weak thinking.
It will feed it.

Chapter 8
Success Is a Sedative

When Good Results Make Fragile Systems Look Smart

Systems don't usually break loudly. They slow down quietly, and success often makes that slowdown harder to see.

Most organizations are trained to inspect failure. Miss the number, and the room fills with questions. Lose the customer, and the postmortem begins. Miss the launch window, and everyone wants to know what broke.

Success creates a different response.

The campaign converts, the customer story lands, the spokesperson drives demand, the dashboard turns green, and the board nods. Everyone exhales, scrutiny drops, and the organization starts treating the result as proof that the system worked exactly as intended.

That's where the danger begins.

Some of the most fragile decisions are the ones that keep working long enough to become invisible. A successful outcome can hide weak reasoning, bad dependencies, reputational risk, customer concentration, timing luck, market tailwinds, channel fragility, or one heroic operator quietly holding the whole thing together with duct tape and caffeine.

That's the sedative effect of success. It calms the room before the system has actually been understood.

Failure forces explanation. Success often gets accepted as one.

A green dashboard tells you what happened, but it doesn't always explain why. It can't tell you whether the result is repeatable, whether the system is healthy, or whether the same decision would survive under different conditions. It only confirms that the outcome worked this time.

In a slower environment, leaders had more room to study successful results over time. They could separate luck from judgment, signal from noise, and momentum from durability.

Compressed time shortens that window. Success turns into doctrine faster, and the thing that worked becomes the thing everyone copies before anyone fully understands what made it effective.

That's how fragile systems scale: through unexamined success.

The organization never announces that it has stopped thinking. It says something much more reasonable:

"This is working."

Because it is working, the room stops looking for the mechanism behind the result. The win becomes the explanation, the metric replaces the reasoning, and success slowly turns from evidence into superstition.

Leaders still need to ask what made the result possible. Which assumptions held? Which conditions mattered? What risk was accepted without being named? What would break if a key person, channel, customer, vendor, model, or market condition changed?

Those questions can feel unnecessary when the numbers are good.

That's exactly why they matter.

Success deserves interrogation too, just without the panic. The goal isn't to punish what worked. It's to understand what came from skill, what came from timing, what came from luck, and what hidden cost may have been absorbed somewhere else before the organization turns one win into doctrine.

When the Win Does Not Travel

This shows up clearly in sales performance. A territory starts working, pipeline improves, win rates climb, and the team assumes the motion is ready to scale.

The result may be real, but hidden conditions are carrying it. One operator has unusual customer context. One channel is warmer than the others. One segment is already primed. One person is quietly bridging gaps the system hasn't actually solved.

From the outside, the motion looks repeatable. Inside, it is fragile in ways the dashboard can't see.

Then the organization scales the playbook across new territories, new reps, and new assumptions while expecting the same result.

The performance doesn't travel.

The win was real, but the organization learned the wrong lesson.

One strong operator can make a messy motion look repeatable. The numbers were real, the customer response was real, and the internal confidence was real. But the outcome depended on judgment that had never been written down, context that had never been transferred, and dozens of small calls one person was quietly making before anyone else saw the problem.

When the organization tried to scale it, the visible playbook stayed the same while the performance dropped away. The company had copied the process without transferring the judgment that made it work.

It didn't have a repeatable motion.

It had a borrowed result.

That's what success can hide: fragility disguised as proof.

Success creates confidence. Confidence reduces inspection. Reduced inspection allows weak assumptions to harden into operating truth. Once that happens, challenging the system gets harder because the evidence appears to be on its side.

The team that questions a failed decision looks responsible. The team that questions a successful outcome can look negative, slow, political, or unwilling to accept the win.

Nobody wants to be the person standing next to a green dashboard saying, "Cool, but what if this is bullshit?"

Sometimes that person is annoying.

Sometimes that person is leading.

That's why success becomes dangerous inside fast-moving systems. It creates momentum and social pressure to keep that momentum going, even when the organization still hasn't figured out what produced it.

Nobody wants to slow down the thing that's working.

Leadership has to protect the organization from obvious failure and hidden fragility. That means inspecting success with the same seriousness applied to failure so the team understands what is creating the result before it tries to reproduce it somewhere else.

The useful question goes beyond whether the motion worked. Leaders need to ask what exactly made it work, which conditions carried the result, and how dependent the outcome was on those conditions.

That changes the conversation. It prevents a good outcome from being mistaken for a healthy system and forces the organization to separate performance from repeatability.

When AI Makes Success Look Cleaner Than It Is

This becomes even more important when AI is involved.

AI can make success look cleaner than it really is. It can produce better-looking work, faster summaries, stronger narratives, and more confident recommendations. When those outputs lead to a positive result, the organization may conclude that the system is working exactly as intended.

Sometimes it is.

Other times, the result still depends on human correction, narrow context, favorable timing, or assumptions that haven't yet been tested under pressure. The machine may have helped while human judgment still carried the outcome.

Leaders who fail to inspect that distinction often scale the wrong thing. They automate the visible parts of success and miss the judgment that made the result possible.

That's how performance turns into exposure.

The first version works. The second works too. Then the third gets rolled out more broadly, and the system finally meets a context it doesn't understand, a customer it wasn't designed for, a constraint nobody named, or an edge case that never appeared during the early win.

By then, the success story has already hardened into internal proof, and internal proof is difficult to unwind.

That's why MACH-10 leaders don't only run postmortems. They also stress-test success.

They ask what may be hiding behind a good result, which parts of the win are repeatable, and which depend on conditions that may not hold elsewhere. They look for fragile dependencies before those dependencies become doctrine.

They don't celebrate less. They understand more.

Keep the momentum, but inspect what's carrying it.

Momentum without understanding is another form of speed without soul. It moves, but it doesn't necessarily learn.

In high-velocity environments, the strongest leaders make success explain itself. They don't accept "it worked" as the full answer.

They ask why.

Then they ask what would put it at risk.

That second question is where judgment lives.

Signals from the Edge

The pattern becomes easier to see in the field.

Win rates improved by double digits, cycle time shortened, and one territory began outperforming the others by enough that the organization treated the result as proof the playbook was ready to scale.

I saw a version of this at PAR Technology while working across sales dynamics, territory execution, and go-to-market strategy. A strong operator, a warmer customer base, and better timing could make the motion look more repeatable than it really was. The numbers were real, but they didn't explain themselves.

The deeper question remained open: which parts of the result were repeatable, and which depended on conditions that might not travel?

The motion was being carried by hidden context. A warmer customer base. A stronger operator. A few unusually helpful relationships. Judgment calls that happened before they ever appeared in the CRM. The dashboard showed the outcome, but it couldn't show what was holding that outcome together.

When the same motion was applied to other territories, the performance didn't travel cleanly. Conversion softened, deals slowed, and the playbook produced weaker results even though it looked the same on paper.

That was the part the dashboard couldn't explain. The team had copied the visible motion and missed the invisible judgment underneath it. The original win depended on timing, customer context, and decisions made by someone close enough to see what the playbook didn't say. Once those conditions changed, the same motion produced a different result.

The organization had scaled the recipe without understanding the chef.

That's where success becomes dangerous. Celebrating the win was reasonable. Scaling the lesson before understanding what created it was the mistake.

A stronger operating model captures the invisible work before expansion begins. It identifies which decisions depended on local judgment, which relationships carried unusual weight, which customer conditions improved the odds, and which parts of the motion can actually be taught, transferred, and repeated.

That gives the organization a chance to separate the real playbook from the temporary advantage. It can preserve what actually drove the result, redesign what depended too heavily on one person, and avoid turning a local success into a company-wide assumption that won't survive outside its original environment.

When those conditions remain invisible, the organization doesn't scale the system. It scales the misunderstanding.

That's how a good result becomes a bad playbook.

The strongest teams don't punish success with bureaucracy. They inspect it with discipline.

They ask what made the win work.

They ask what would make it stop working.

Then they scale only the parts that can survive outside the original conditions.

Mindset Shift to Adopt

Stop treating success as proof that the system is healthy.

Start treating it as evidence that still needs inspection.

Tools to Try This Week

Pick one recent win and run a success premortem.

Identify what made the result work. Name the conditions that were required, list the assumptions the team accepted without testing, and separate the parts of the win that are repeatable from the parts that may have been situational.

Then look for what the dashboard left out. Which person supplied judgment the process didn't capture? Which relationship made the path easier? Which customer, timing advantage, market condition, or operational workaround helped carry the result? Ask what would change if any of those conditions disappeared.

Don't let a good outcome become a lazy explanation. Use AI to surface possible hidden dependencies, alternative explanations, and failure conditions, but don't let a clean explanation replace real inspection.

The goal is to understand what can travel. Capture the decisions, context, and judgment that made the win possible before turning it into a playbook. Strengthen the parts that depend too heavily on one person, and test the motion under conditions that are less favorable than the original win.

Validate the lesson before you institutionalize it, because once a weak explanation becomes the playbook, every new team inherits the same misunderstanding.

If the team scales the wrong lesson, success becomes the thing that teaches the system badly.

Do This Now

Choose one recent success. Pick something your team is celebrating right now.

Write down why you believe it worked. Do not settle for "the campaign worked," "the customer liked it," or "the numbers improved." Name the mechanism.

Separate repeatable from situational. Identify which parts of the win can travel and which parts may depend on timing, context, channel, customer, operator, or market condition.

Name one fragile dependency. Identify one thing that would break the model if it changed.

Ask the second question. What would make this stop working?

Adjust the playbook. Fix the lesson before the win turns into doctrine.

Key Takeaways

Failure creates attention, while success often removes it. When the numbers are bad, teams investigate. When the numbers are good, they're more likely to move on.

A good outcome doesn't always mean the system is healthy. It may only mean the conditions were favorable enough to hide what was weak underneath.

The result may be real while the lesson is wrong. A team can accurately measure what happened and still misunderstand what caused it.

Momentum without understanding is speed without soul. The organization keeps moving, but it doesn't necessarily become more capable of repeating or protecting the result.

MACH-10 leaders inspect wins before they scale them. They preserve the momentum while identifying which parts of the success can actually survive outside the conditions that created it.

Leaders shouldn't ask only whether a win can be repeated. They should ask what has to remain true for that repetition to be possible.

MACH-10 Leadership Moves

Run a stress test on one recent success. Don't wait for the failure to start asking better questions.

Separate repeatable advantage from temporary performance. Identify what can scale and what only worked under specific conditions.

Identify one hidden dependency inside a current win. Look for the operator, channel, customer, timing, or assumption carrying more weight than the system admits.

Ask what would break if the context changed. Make the win defend itself before it becomes doctrine.

Don't scale success until you understand what made it work. Scale the system, not the superstition.

Celebrate the win.
Then make it explain itself.

Chapter 9
The Automation Trap

When Speed Scales the Wrong Work

Automation usually enters the room wearing a name tag that says efficiency.

It promises less work, faster output, cleaner processes, and a dashboard that looks like someone finally did something useful. Sometimes that's exactly what it delivers. Other times, it simply moves the mess to a part of the system nobody is measuring yet.

That's the part leaders have to slow down long enough to see. The visible work may be shrinking, but the real cost hasn't necessarily disappeared. It may have moved into review, correction, escalation, customer confusion, or the quiet human effort required to keep the automated process from drifting into nonsense with a confident smile.

The logic is easy to follow. If AI can draft the response, why not let it handle the workflow? If it can generate the recommendation, why not route more of the process through it? If it reduces effort at one point in the system, why not remove the human layer entirely?

Automation doesn't erase complexity simply because it removes visible effort. More often, it relocates the complexity and makes the new cost harder to see.

At one company, my CEO asked me to examine an AI workflow that was moving faster but creating too much review behind it. On the surface, the system looked more efficient. Output had increased, cycle time had improved, and the automated layer appeared to be doing its job.

The work, however, hadn't disappeared. It had shifted into a review layer that was beginning to consume the time of the same people the system was supposed to help. We had reduced effort at the front of the workflow and quietly recreated it downstream.

The answer wasn't to abandon the automation. We moved from universal review to exception review, allowing routine outputs to keep moving while human judgment stayed focused on cases where risk, ambiguity, or consequence actually mattered.

That changed the economics of the workflow. Oversight became proportional to consequence instead of being applied evenly to everything. The speed remained, but human attention stopped getting spent on predictable work and became available for the cases most likely to create downstream damage.

That's the Automation Trap.

It looks like efficiency because work happens faster, output increases, and the apparent cost of producing something goes down.

Then the hidden bill arrives.

Coordination increases. Corrections multiply. Oversight migrates downstream. Decisions start moving faster than the organization can understand or absorb them.

The work didn't disappear.

The cost moved.

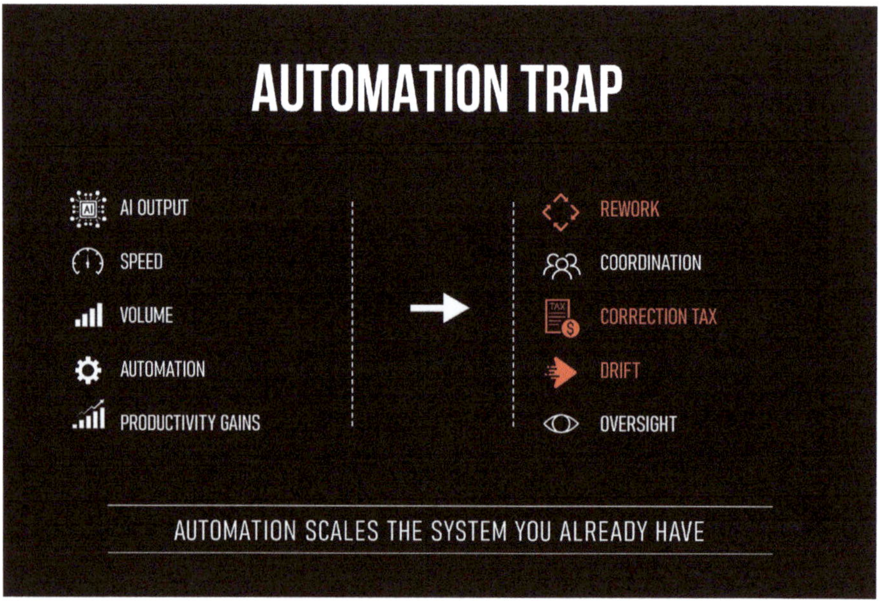

FIG 9.1 – Automation Trap. Automation scales the system you already have.

When Efficiency Moves the Cost

Sometimes the cost moves into coordination.

I've seen teams "save time" on generation and spend twice as long cleaning up the downstream issues the automation created. I was part of one of those teams.

We automated part of a customer-facing workflow, and the first few weeks looked like a clean win. Response times dropped, more requests moved through the queue, and the dashboard improved. It was easy to believe the process had taken cost out of the system.

The simple cases were faster, but the messy ones got stranger. Customers came back with follow-up questions the workflow hadn't resolved. Managers spent more time reviewing exceptions, while support teams untangled issues that used to be caught earlier because a person had been close enough to see what was actually happening.

The dashboard said the front end was faster.

The team felt the back end getting heavier.

That's the Correction Tax.

Every shortcut AI creates upstream sends a bill somewhere downstream unless the workflow is designed to catch it. The expense rarely shows up in the first productivity metric. It accumulates as rework, second-guessing, exception handling, and time spent unwinding decisions that looked efficient when they were made.

No one tracks it until the process starts feeling heavier.

Errors begin repeating at scale. Volume increases, but consistency doesn't hold. The operation looks productive even as the value it creates becomes less predictable.

The pattern is simple: teams optimize for speed at the point of creation instead of across the full workflow.

AI makes it easier to produce more, but it doesn't guarantee that what gets produced will survive contact with reality. A gain in one step can create additional effort somewhere else, often in places the original efficiency calculation never considered.

That's why the cost conversation around AI is usually incomplete. At small scale, it looks cheaper. At system scale, it depends.

Integration still takes time. Exceptions still need handling. Errors still need correction. Processes still change.

None of that disappears.

It just finds a new place to hide.

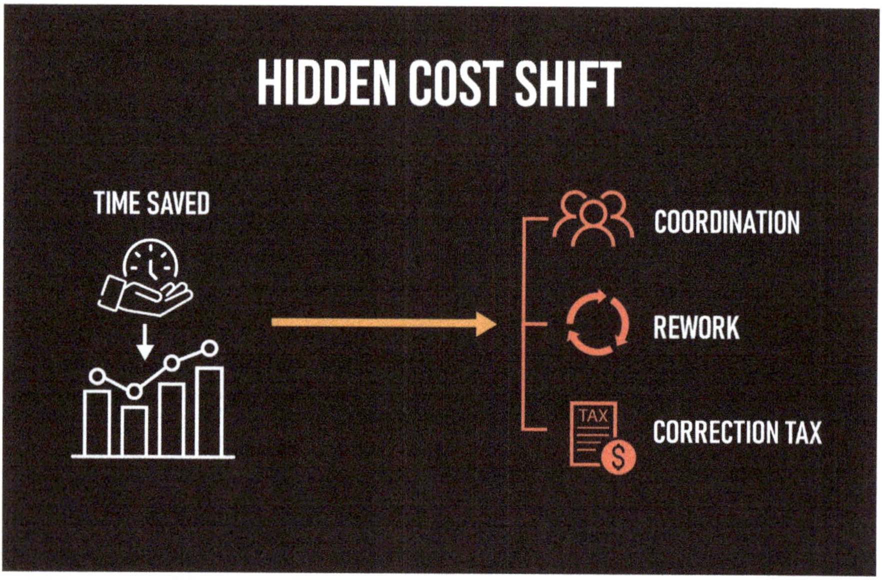

FIG 9.2 – Hidden Cost Shift™. AI often saves time at the point of creation by shifting cost into downstream coordination, correction, and oversight.

The Hidden Cost Shift

When AI reduces visible effort upstream, downstream correction, coordination, and oversight can quietly increase.

The more invisible the cost becomes, the more dangerous it is.

Errors behave differently in automated systems. A human mistake is usually contained. It has context. It can be traced and corrected.

An AI-driven mistake moves.

It can pass through multiple steps before anyone notices. It can be reused, referenced, routed, embedded, and acted on. By the time it surfaces, it's harder to unwind.

The error is bad enough.

The real damage starts when the organization carries it forward.

That's why automation requires a system view. A gain in one step isn't always a gain for the whole organization. If one team saves an hour and another team spends three hours cleaning up the result, the organization didn't get faster.

It shifted the burden.

And that burden became someone else's operating model.

That's the hidden cost shift. The work did not vanish; it changed shape, location, and owner.

Augment Before You Replace

There's also a tendency to move too quickly from augmentation to replacement.

Once AI can handle a task, the next step can feel obvious: remove the human, reduce cost, and increase speed. Sometimes that works. Often, it doesn't.

AI doesn't eliminate work. It redistributes it.

What remains is oversight, correction, exception handling, and decision-making in the areas where the automation is less certain. When teams remove people too early, they also remove the context needed to manage those edges.

The process may become faster, but it also becomes harder to control.

A more stable approach is to augment first. Use AI to increase capability, watch how the workflow behaves, and identify where it performs well and where it struggles. Find the edges before those edges become customer-facing problems.

Some human steps can be removed. Others should remain because they carry judgment, context, or accountability the system still needs.

That decision has to be made from inside the workflow, based on how the system actually behaves, not from outside it based on what the automation appears capable of doing.

The Four Automation Tests

Before automating anything, run it through four tests.

Is the outcome reversible if it's wrong? Is the context stable enough for the pattern to hold? Will the cost of an error become visible before it compounds? Is someone clearly accountable for the decision?

If you can't answer yes to those questions, you're not reducing risk. You're scaling it.

This is where judgment becomes the constraint. Not every process benefits from automation, and not every decision should be accelerated. Some areas depend on context that's difficult to encode, while others carry consequences that are too costly or too difficult to reverse.

Judgment determines whether AI has a seat at the table, how much influence it should have, and where human control has to remain.

AI doesn't reduce the need for judgment.

It raises the cost of operating without it.

Ownership Cannot Be Automated

As AI becomes more deeply embedded in operations, the question of responsibility gets harder to ignore. The system may generate the recommendation, route the work, or influence the decision, but it can't absorb accountability when the outcome goes wrong. That responsibility still belongs to the people who designed the process, accepted the output, and allowed it to move.

Most conversations about responsible AI focus on bias and compliance.

But the operational issue is ownership.

When decisions are influenced by AI, responsibility doesn't move into the system. It stays with the leader.

As systems move faster, decisions compound faster. Errors travel farther. Outcomes materialize sooner. The margin for correction shrinks.

Speed can stay in the system.

The dangerous part is letting mistakes outrun understanding.

Signals from the Edge

The Automation Trap usually looks harmless at first. In a restaurant technology environment, a team used AI to speed up reporting and recommendation prep. Turnaround dropped 52%. Reports were cleaner. Summaries were sharper. The early numbers made the workflow look better.

But speed exposed a gap we had not closed. Ownership. One recommendation went to a key operator with a polished but wrong pricing assumption. Everyone could explain the workflow, but no one could explain the decision. The cleanup landed on the CSM, who had to call the customer and defend a number she did not create. That call took three hours. The trust took longer to rebuild.

Automation made us faster. It didn't make us accountable. The gap wasn't in the model. It was in the ownership around it.

That distinction matters because accountability can't be retrofitted after the output has already reached the customer. By then, the organization is repairing trust. Clear ownership has to exist before the recommendation moves, so someone is responsible for validating the assumption, accepting the risk, and stopping the workflow when the answer looks polished but the reasoning underneath it is weak.

Mindset Shift to Adopt

Stop assuming AI reduces effort.

Start asking where the effort moves.

Tools to Try This Week

Review one AI-driven workflow end to end.

Map what happens before it, after it, and because of it.

Identify where correction, coordination, cleanup, review, or escalation shows up downstream.

Ask who owns the outcome when the workflow produces something wrong.

Compare the promised efficiency gain against the actual correction cost.

Replace one full automation with an augmented approach and compare the total system cost, not just the visible time saved.

Do This Now

Pick one automated workflow. Choose something your team has already automated or wants to automate.

Map the work before AI enters the process. Identify what humans used to notice, catch, or correct.

Map the work after AI creates output. Look for review, correction, routing, escalation, exception handling, or customer cleanup.

Find where the cost moved. Identify where effort increased downstream even if the front end got faster.

Identify the hidden cleanup. Name the review, follow-up, escalation, or correction work missing from the original efficiency claim.

Name the owner. Make one person accountable for the final decision, not just the workflow.

Ask the real cost question. Is the process actually cheaper, or did the cost simply move?

Key Takeaways

AI doesn't eliminate work. It redistributes it.

Automation before understanding scales complexity.

The Correction Tax shows up as rework, review, exception handling, and downstream confusion.

Visible speed gains can hide system-level cost.

A gain at the point of creation isn't always a gain for the system.

MACH-10 Leadership Moves

Review one automated workflow from end to end. Don't stop measuring where the AI output appears.

Identify where AI has moved effort downstream. Look for correction, coordination, review, escalation, or exception handling.

Define the verification point. Decide where AI output can move freely and where human review must interrupt the loop before the system acts.

Measure total system cost. Track more than time saved at the point of creation.

AI doesn't eliminate work.
It decides where it shows up.

Chapter 10

Trust Becomes a Decision

Where to Rely on the System and Where to Apply Judgment

Trust used to feel like something you earned at the end.

Review the work. Check the assumptions. Approve the decision. Then let it move.

That model breaks when the work is already moving before anyone has time to inspect it.

Inputs arrive continuously. Decisions happen in motion. Outputs move through the system before anyone can inspect every detail. The organization still needs trust, but it can no longer treat trust as a final approval step. By the time a traditional review process catches up, the output may already have shaped a customer conversation, influenced a recommendation, redirected a workflow, or created work somewhere else.

In an AI-native system, trust isn't a warm feeling.

It's a routing decision with consequences.

That's where leaders get uncomfortable. The instinct is to verify more. More review. More confirmation. More checkpoints. More time spent slowing the system down just enough to feel in control. It looks responsible because it resembles diligence.

Applied evenly, though, that instinct creates its own problem. The system carries too much hesitation where speed matters and too little scrutiny where failure is expensive.

That's the trust problem now.

Not whether to trust.

Where to trust.

Some outputs can move immediately. Others need sampling, deeper interrogation, or a human owner close to the call. A low-risk meeting summary can move quickly because the cost of a missed nuance is usually small and easy to correct. A pricing recommendation is different. It may look just as polished, but the consequences touch revenue, positioning, customer expectations, and competitive response.

A customer-facing automation carries more weight still. If it gets the answer wrong, the mistake doesn't stay inside the team. It reaches the market. A compliance-sensitive decision belongs in another category entirely. Casual confidence has no place there, no matter how clean the output looks.

Same technology.

Different trust posture.

Leaders who operate effectively at speed don't try to inspect everything. They decide which outputs can move freely, which deserve sampling, and which should stop until someone owns the consequences. If nobody manages that balance deliberately, the organization eventually lands in the worst version of both worlds: too much trust where the cost of being wrong is high, and too little where review adds almost nothing.

AI makes that calibration harder because it compresses many of the signals leaders once used to judge quality. The writing may be clean. The summary may be useful. The recommendation may be directionally right. None of those qualities tells you how much authority the output should have.

Trust has to follow the consequence, not the polish.

The question isn't how much trust the system deserves.

The question is where trust belongs.

Trust Belongs to the Use Case

Trust isn't granted to the tool.

Trust is assigned to the use case.

That makes trust a routing decision, not a belief system.

A model can produce useful work in one context and risky work in another. A summary may be fine for orientation and dangerous as the basis for a high-consequence decision. A recommendation can be valuable as input and reckless as instruction.

"Do we trust AI?" is the wrong starting point.

That's how you end up having an I, Robot conversation in a spreadsheet meeting.

The better question is smaller and much more useful:

What are we trusting this output to do?

Are we trusting it to inform, decide, approve, reject, explain, or act?

Those are very different permissions. Informing a person may require little intervention. Acting on a customer account, changing a price, or triggering a compliance-sensitive workflow may require a named owner, a threshold, and a clear path for correction.

Trust, in this environment, has to be directed. You have to stay close enough to recognize when something stops fitting the reality around it.

Distance Hides Signal

As systems scale, leaders get further from the details and rely more heavily on summaries, dashboards, and compressed outputs. That creates efficiency.

It also creates distance.

And distance hides signal.

The further leaders move from the original work, the easier it becomes to confuse compression with understanding. A dashboard can tell you what changed without telling you why. A summary can remove the exact piece of context that would have made the recommendation look questionable.

Leaders who stay effective at speed don't try to review everything. They sample. They stay close to the edges of the system. They look for inconsistencies, strange exceptions, customer behavior that doesn't fit the story, and places where an output feels slightly too clean for the situation it is describing.

They pay attention to the things that feel off before they can fully explain why.

That's usually where the useful signal is hiding.

Trust Has to Be Steered

Trust isn't something you set once and walk away from.

A workflow that can run with light oversight in one environment may need tighter controls in another. A customer-facing output deserves a different posture than an internal draft. A low-risk summary can move quickly, while a decision with financial, legal, reputational, or customer consequences needs stronger checks and a clearer owner.

Teams get into trouble when yesterday's permission quietly becomes tomorrow's default. A system performs well, confidence rises, scrutiny fades, and suddenly a workflow designed for one set of conditions is operating under another. Nobody made a new trust decision. The old one simply survived longer than the context that justified it.

That is why trust has to be steered rather than granted.

Ownership Carries the Trust

Ownership becomes more important in this environment because trust doesn't carry accountability.

Every decision still needs an owner, and every output that influences action needs someone responsible for how it's used. When that isn't clear, the system continues to move while responsibility starts evaporating around it.

I've been in rooms where everyone agreed something was off, and nobody took the first step to fix it because ownership was diffuse. The delay wasn't about awareness. It was about responsibility.

Slower correction is only part of the cost. People start wondering whether the output made the call, whether the room made the call, or whether the call somehow emerged from the fog wearing a nametag.

That ambiguity burns time.

It also gives everyone a place to hide.

Clear ownership removes the hiding place.

Leaders who handle this well make responsibility explicit before the output starts moving. They define who can use the recommendation, who can override it, who watches the result, and who acts when reality pushes back.

The system can generate recommendations, rank options, summarize evidence, and surface patterns.

It cannot own the consequence.

That responsibility doesn't move.

Uncertainty is part of the operating environment now. Strong leaders stop waiting for it to disappear and build controls that let the organization move inside it. That means knowing where the system can run, where friction earns its keep, where speed creates advantage, and where a human has to remain close enough to challenge the call.

Trust is not a blanket decision.

It's a control system.

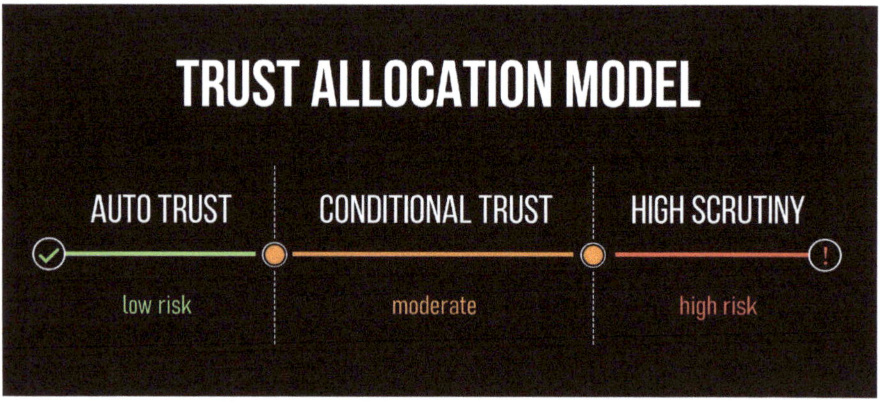

FIG 10.1 – Trust Allocation Model™. High-speed systems require selective trust, not uniform validation.

Signals from the Edge

I've watched product teams fall in love with AI-generated summaries because the outputs were clean, fast, and easier to consume than the raw mess underneath. In one workflow, weekly customer and performance summaries that used to take days came together in hours. Customer feedback, performance data, support notes, sales input, and usage trends were suddenly pulled into something that looked organized and useful.

Then someone asked why one customer pattern kept showing up so strongly.

Certain patterns were overrepresented. Others were missed. The summaries reflected what the system was most confident about, not always what mattered most.

The output created a sense of certainty the team hadn't earned yet.

That is the danger of clean synthesis: it can make uncertainty look resolved.

By the time the team went back to the underlying data, the summary had already shaped the conversation. The team had spent its energy debating the pattern the summary made visible while the more important signal sat underneath it.

A summary can be useful and still be too dangerous to treat as the truth. The better teams treat AI outputs as starting points. They question them, test the assumptions, and validate where the decision carries risk.

They use the output without surrendering to it.

They keep it on probation.

Mindset Shift to Adopt

Stop trying to verify everything.

Start deciding where verification actually matters.

Tools to Try This Week

Sample outputs regularly instead of reviewing everything.

Define the trust boundary for decisions that rely on AI-generated inputs.

Add one check where errors would actually carry cost, not where review simply feels safe.

Do This Now

Choose one AI-assisted workflow. Pick a workflow where summaries, recommendations, drafts, or analyses are already moving through the system.

Decide what the output is trusted to do. Is it for orientation, recommendation, routing, customer response, or decision support?

Sample the output. Compare a few outputs against the underlying data, customer reality, or operational context.

Adjust the trust boundary. Move quickly where the risk is low, and add scrutiny where the cost of error is real.

Key Takeaways

Trust becomes selective as speed increases.

Trust belongs to the use case, not the tool.

Outputs are inputs, not decisions.

MACH-10 Leadership Moves

Define where your team can move without additional validation. Give low-risk outputs room to move.

Identify where errors would be costly. Reinforce those points with ownership, sampling, or review.

Clarify who owns AI-influenced decisions. Do not let the system become the decision-maker by default.

Review a sample of outputs. Understand how the system is actually behaving, not just how clean the output looks.

You don't need to control everything.
You need to know where to look.

Chapter 11

Your Organization Is Slowing You Down

Speed Breaks Systems Designed for Control

Responsibility doesn't scale automatically, and structure is never neutral. It either helps decisions travel or quietly taxes them.

Most organizations were built for a slower clock. They were designed for control, predictability, and a pace where choices could pass through several layers without losing much value. That model worked because there was more space between action and consequence, giving leaders time to think, align, revise, and adjust before the impact arrived.

That space has largely disappeared.

What once felt like sound structure can now become its own source of delay. Handoffs create gaps, approval chains add time, and decisions wait for sign-offs that rarely change the answer. The organization may be functioning exactly as designed while the design itself no longer matches the environment around it.

That's the real problem.

Many organizations are dealing with a permission architecture problem. The constraint isn't always how quickly a leader can think. It's whether the organization allows a good decision to travel without dragging it through process that adds ceremony but little value.

An AI-native organization is defined by how work flows through the business. Decisions happen closer to the point of action, responsibility stays visible, feedback loops remain short, and teams keep operating instead of waiting for the next formal stage to begin.

Traditional organizations tend to organize work sequentially: input, analysis, approval, execution, review. Each phase is controlled, every transition is managed, and each handoff adds a little more delay.

For years, that was manageable.

Now the delay compounds.

By the time the work reaches the end of the sequence, the conditions behind it may have already changed. The plan can still look perfectly coherent inside the company while failing the moment it meets the market. That has less to do with effort or intelligence than with the architecture surrounding the work.

AI-native organizations rely more heavily on flow. Work doesn't stop at every boundary. Input turns into a decision, the decision turns into action, feedback arrives, and the next adjustment begins while the conditions are still relevant.

Better tools can expose the problem. Leadership still has to fix it.

This is where plenty of transformations go sideways. Teams accelerate individual steps while leaving the unnecessary ones intact. They make the old process faster without asking whether parts of it should exist at all.

That's like putting a better engine in a car with the parking brake on. The dashboard improves, but the car is still fighting itself.

Hierarchy can still serve a purpose. The real constraint is latency.

Every additional layer creates another chance for delay and distortion. Information travels upward, decisions come back down, and some amount of context gets shaved off in both directions. In a slower environment, that trade-off was tolerable. At higher speed, it starts costing the organization more than the control it provides.

I've watched good decisions weaken as they traveled up and down an organization. By the time they returned, they were safer, cleaner, and much easier to defend.

They were also less useful.

AI-native organizations reduce the number of people a time-sensitive decision has to pass through. Authority sits closer to the work, while leaders spend less time serving as approval gates and more time defining direction, boundaries, and consequences.

That changes the leadership job. Leaders don't need to make every call themselves. They need to build an environment where the right calls can happen quickly without sacrificing responsibility, judgment, or the information that made the decision useful in the first place.

Traditional operating models also tend to batch work. Information is collected, decisions happen at designated points, execution proceeds in cycles, and feedback gets reviewed after the fact. That can be efficient when conditions are stable. When they're changing quickly, the waiting becomes expensive.

An AI-native model keeps more of that activity continuous. Information gets processed as it arrives, teams act once they have enough confidence, and feedback starts influencing the next decision almost immediately.

As the distance between thinking and doing shrinks, the organization gets more chances to learn while the learning can still change the outcome.

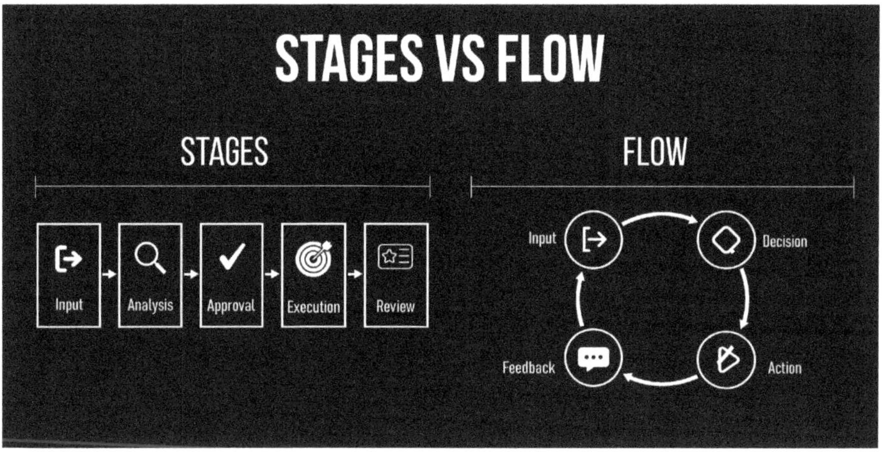

FIG 11.1 – Stages vs. Flow Systems. High-speed organizations operate through continuous flow instead of sequential stages.

At Audivi, the speed gap showed up inside internal workflows and reporting.

The work was ordinary, which made it a good test. There wasn't a giant transformation program or dramatic rollout. It was simply the kind of routine that quietly determines whether a small team stays ahead of the business or spends the week catching up to itself.

AI helped immediately.

Work that used to take most of a morning could now reach a usable first version in under fifteen minutes. Inputs were easier to gather, context was easier to summarize, and the team could see more of the picture sooner.

The report wasn't the finish line. It existed to support a decision.

Once the output was ready, the same operational questions were still waiting for us. Who owned the next move? What needed attention now? What could wait? Which cases required a human read before they moved? Which signals were only noise with better formatting? Most important, what decision was the report supposed to make easier?

Those questions forced the issue. The workflow had improved, but the decision loop around it still needed work. We could prepare the report faster than before, yet the operating model still lacked a clear way to turn that output into action.

That distinction mattered because faster preparation could create the illusion that the whole system had improved. In reality, the gain only became useful when ownership, review, and action moved with it. Otherwise, the report arrived sooner and simply waited longer in a different part of the workflow.

So we tightened the loop.

The first change was ownership. The report couldn't land in a shared space and hope someone would turn it into movement. One person needed to own the next call. Others could add context, challenge the interpretation, or flag risk, but the decision needed a clear home.

The second change was the review path. Some review improved judgment. Other review slowed an output that was already clear enough to use. We kept review where it could change the decision and removed it where it only added comfort.

The third change was the threshold. The team needed to know what made the report actionable. If the output met the agreed quality bar and didn't surface an exception, the owner could move it forward. When something looked off, uncertain, or high-risk, it went back through human judgment before action.

Routine work could keep moving without unnecessary friction, while cases with greater consequence still received the attention they deserved. The system became faster because the team knew where judgment belonged, not because every step had been stripped away.

AI organized the inputs, summarized the context, and helped surface the parts that needed attention. The owner decided what mattered. The review path handled exceptions instead of treating every report like an exception. Feedback came back sooner because the decision reached the work sooner.

The difference showed up in small moments.

A report didn't sit while people figured out who was supposed to respond. A low-risk follow-up didn't need the same level of review as a sensitive call. An exception had a defined path instead of becoming another thread of interpretation. A miss surfaced while the context was still fresh enough to fix. Real operating change happens when ordinary decisions stop losing time at the boundaries between information, ownership, judgment, and action.

That's where the stages-versus-flow distinction became practical.

In the stage version, the work moved from input to report to review to discussion to decision to action. Each step had a reason to exist. The problem was the waiting between them, especially when nobody knew whether the next step was adding value or simply preserving the sequence.

In the flow version, the report stayed connected to the decision it served. Input became signal, signal moved to an owner, the owner applied judgment, and action created feedback. That feedback improved the next report and the next decision.

AI shortened the distance between information and usable output. Leadership had to shorten the distance between usable output and action.

We still had to decide who owned the call, what level of judgment was required, and how quickly feedback could return to the loop. Those decisions became the real operating model because they determined whether the speed created by AI translated into movement or simply produced faster waiting.

That was the Audivi lesson.

The tool could prepare the work faster. The operating model determined whether the decision moved.

That's where structure stopped being background and became the work.

AI wouldn't repair structural problems on its own. It would expose them.

Faster tools inside a slow structure created uneven performance. Some parts of the organization moved quickly while others remained constrained. The gaps between them widened, coordination became harder, and the system grew less predictable.

AI amplified the structure it operated within. When the organization was built around permission, speed kept running into approval. When ownership and decision paths were clear, the same tools created movement instead of more waiting.

Structure earned its place when it helped work move.

Mindset Shift to Adopt

Stop trying to make a slow structure look faster.

Start designing the organization so work can actually move.

Tools to Try This Week

Move authority closer to the work for one decision that currently involves too many stakeholders.

Remove one approval step that doesn't change the outcome.

Do This Now

Pick one slow workflow. Choose a workflow that feels heavier than it should.

Map every handoff. Write down each approval, review, handoff, and decision point.

Find one low-value step. Identify one step that adds time without changing the outcome.

Remove it or time-box it. Do that this week, not someday after six more sync meetings.

Key Takeaways

Structure determines speed.

AI amplifies whatever structure already exists.

If the organization is built around permission, speed will keep asking for approval.

MACH-10 Leadership Moves

Identify where structure is slowing decisions down. Look for the places work waits for permission.

Push one decision closer to the work. Move authority nearer to the people with the best context.

Remove one approval layer from a decision path. Cut the step that adds comfort but not quality.

Replace one permission step with a decision rule. Let the team move when the rule is met instead of waiting for approval.

The system does not need to be perfect.
It needs to stop waiting.

Chapter 12

Strategy Is a System, Not a Plan

Why Learning Speed Now Determines Advantage

There was a time when strategy was something you could finish.

You defined a direction, analyzed the market, aligned the organization, built the roadmap, and moved into execution. The process felt controlled because the environment usually gave the plan enough time and breathing room to matter.

That assumption is gone.

A strategy can still look finished while the market is already moving around it, which is what makes drift so difficult to catch.

The environment now changes faster than the planning cycles most organizations still rely on. Inputs shift continuously, competitors adjust in real time, and customers respond immediately. By the time a strategy is fully defined, the conditions that shaped it may already be changing.

Nothing breaks right away, and that is part of the problem.

Execution continues. Work moves forward. Teams remain aligned, and the operating model looks healthy from the inside. They are simply aligned to a version of reality that is starting to expire.

I've seen teams execute a strategy exactly as designed and still lose position because the environment changed while they were aligning to it. Nothing was obviously wrong with the plan. The analysis was sound, the execution was disciplined, and the strategy held together internally.

It had simply lost relevance externally.

The team was still executing the plan while the market had already moved on from the premise.

Leaders still need strategy. The failure comes from treating it as a document instead of a system that has to stay connected to changing conditions.

The real question is whether strategy can adapt quickly enough to remain useful.

In a slower environment, strategy could behave like a plan. Leaders defined the destination, sequenced the work, aligned the teams, and executed the roadmap.

Now it has to behave more like a living system. It needs to absorb new information, guide decisions, adjust direction, and keep teams moving without forcing the entire organization to stop and realign every time something changes.

Most organizations haven't made that shift yet.

They still treat strategy as a phase: quarterly planning, annual reviews, roadmap sessions, and executive offsites with catering, slides, and enough abstract nouns to tranquilize a water buffalo.

Those rituals can still create value, but they can't carry the entire strategy system. A planning cycle is too slow to be the only place strategy happens.

By the time the deck is polished, the market has already started editing it.

When the Plan Starts Aging Out

The gap between defining strategy and acting on it becomes a source of risk. The longer that gap stays open, the more drift accumulates. By the time execution catches up, some of the assumptions that shaped the plan may no longer hold.

The plan may still make sense on paper.

The market doesn't grade on paper.

A strategy is only useful if the organization can keep it alive after the planning meeting ends.

AI-native strategy closes that distance by tightening the connection between new information, decisions, and action. Direction still matters. Alignment still matters. Trade-offs still matter. But they can't sit frozen inside a planning artifact while reality keeps changing around them.

Strategy stays relevant when it operates as a continuous loop between evidence, decisions, execution, and learning. New information doesn't mean abandoning direction every week. It means having a way to test whether the assumptions supporting that direction still deserve to survive.

Without that loop, organizations confuse consistency with relevance. They keep executing because the plan still looks coherent even as the evidence changes.

The roadmap remains beautifully aligned while reality quietly removes itself from the meeting invite.

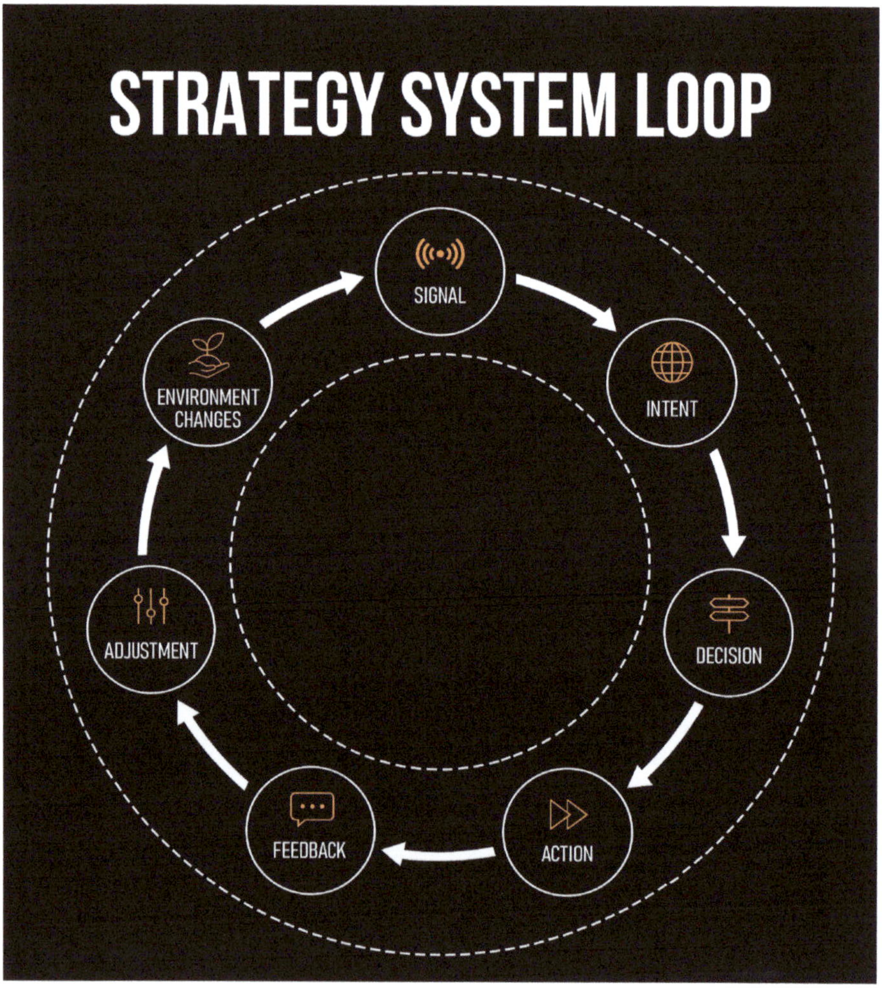

FIG 12.1 – Strategy System Loop. Strategy stays alive when signal, intent, decision, action, and feedback remain connected.

Strategy Has to Stay Alive

That means decisions are no longer simply downstream from strategy. They are how strategy gets expressed, tested, corrected, and reinforced over time. Every meaningful decision either strengthens the direction or gradually dilutes it.

That's why strategy can't be separated from decision speed.

If decisions move slowly, strategy becomes stale. If they move without direction, strategy turns into noise. Leaders have to manage both at once by providing enough direction to keep the system coherent and enough movement to keep the strategy alive.

The Leader Becomes the Intent Engine

AI can process information faster. It can synthesize data, identify patterns, generate options, and summarize trade-offs.

It still can't decide what matters, own the trade-off, or understand what kind of company you're trying to become. That remains leadership work.

The leader defines intent, and that intent gives teams a way to act without waiting for every answer to travel back up the chain. The faster conditions change, the more important that becomes. People need enough guidance to respond on their own without turning every adjustment into another approval cycle.

Clear intent reduces hesitation. Weak intent creates noise because teams still move, but they do so through interpretation rather than shared direction.

When intent is clear, teams can make faster decisions without losing coherence. They understand the priorities, constraints, trade-offs, and boundaries well enough to respond as conditions change. They don't have to stop each time reality shifts because the strategy is already embedded in how decisions get made.

That is the difference between a plan and a system.

A plan tells people what to do. A strategy system helps them decide what to do when the plan no longer matches reality.

Strategy is no longer something leaders finalize and hand off to execution. It is something they keep running through decisions, feedback, and adjustment. New information changes what the organization knows, but it shouldn't force everyone back to the starting line every time the market does something inconvenient.

Signals enter the system, intent gives them meaning, decisions create movement, and feedback improves the next decision. The value comes from keeping direction and learning connected while the environment continues to move.

Without that connection, organizations split in two. Strategy becomes a story executives tell, while execution becomes a collection of local decisions made under pressure. Everyone stays busy, but the work gradually stops reinforcing the direction.

That is how companies drift.

The strongest leaders prevent that drift by making strategy operational. They clarify what matters, define trade-offs, move authority closer to the work, and shorten the distance between insight and action. They use AI to accelerate sensemaking while keeping human judgment responsible for intent.

That is what makes strategy durable now.

Its strength doesn't come from remaining unchanged. It comes from being able to change without losing itself.

Signals from the Edge

Earlier in my career, I was on a team that identified a shift in customer behavior early and still lost the moment because the strategy system couldn't convert the signal into action quickly enough.

The signal was clear. Usage patterns were changing, and the data pointed directly to an opportunity to reposition a core feature.

The decision didn't move with it.

Instead, it passed through additional analysis, broader review, and multiple approval layers. By the time the change was approved and released, a competitor had already acted on the same signal.

The advantage didn't come from seeing something we had missed. It came from conversion speed: turning the same information into action before the slower system finished processing it.

That kind of delay is easy to rationalize from inside the organization. Each request for more evidence sounds responsible, and each review appears small on its own. Together, they create a system where everyone can see the opportunity and nobody can reach it before the window closes.

The stronger pattern was more connected.

A smaller team operated with clearer intent. The team understood the strategic priority, knew the relevant trade-offs, and had enough authority to act. When a partial signal appeared, they didn't wait for a full planning cycle.

They made a bounded move, tested the assumption, and adjusted quickly. The decision carried enough risk to produce useful evidence without putting the entire strategy on a roulette wheel and asking finance to blow on the dice.

That approach also changed the role of certainty. The team didn't need complete confidence before moving because the move itself was designed to generate better information. Action became part of the strategy process rather than something that happened only after the strategy was considered finished.

The strategy improved because the system stayed in contact with reality.

The initial decision wasn't perfect. It was connected, and that made the difference.

Mindset Shift to Adopt

Stop treating strategy as something you finalize.

Start treating it as something your system runs every day.

Tools to Try This Week

Take one active strategic initiative and map how strategy turns into decisions.

Identify where signals enter the system.

Identify who has authority to act on those signals.

Find where decisions are slowing down because intent is unclear.

Use AI to summarize market, customer, and performance signals. Keep leadership accountable for what those signals mean.

Replace one planning discussion with a decision that tests the strategy in the real world.

Do This Now

Pick one strategic priority. Choose something your team is already working on.

Name the decision it should change. Write down the decision this strategy should affect this week.

Clarify the trade-off. Decide what matters most right now and what you're willing to say no to.

Move one action forward. Do it before the next planning conversation.

Key Takeaways

Strategy has to operate inside the system, not above it.

A plan loses value when the environment moves faster than the planning cycle.

Clear intent lets teams move without constant realignment.

AI can accelerate sensemaking, but leaders still own direction.

MACH-10 Leadership Moves

Clarify one strategic intent. Give your team a direction they can use to make faster decisions.

Find where strategy is stuck. Identify one place where strategy is trapped inside a planning process.

Create one strategy trigger. Define the signal that should cause the strategy to adjust before the next planning cycle.

Replace one sync meeting with a strategy test. Learn from reality, not another readout.

Review whether decisions are reinforcing the strategy. Look for where the work is strengthening the direction or diluting it.

The system does not reward the best plan.
It rewards the strategy that keeps learning.

Chapter 13

Execution Drives Decision Velocity

Learning Speed Determines Performance

Execution is where strategy stops pretending.

A plan can sound brilliant in the room. The slides can be clean. The logic can hold. Everyone can nod like the future has just been professionally formatted.

Then reality gets involved.

In a system that keeps moving, execution isn't downstream from strategy. It's how strategy learns. The plan becomes useful only after reality gets a vote, because what the team does, how quickly it does it, and how it responds next start shaping the direction in real time.

Strategy is no longer something you complete before you act. It gets refined through action, where assumptions meet customers, constraints, timing, and consequences the planning session couldn't fully predict.

Most teams still treat execution as something that follows alignment. They wait for clarity before moving and want the plan to feel complete before putting it into motion. That feels responsible, but it creates a dangerous gap between what the organization believes and what the environment is already showing it.

The organization ends up catching up to reality.

Execution Creates Clarity

High-performing teams operate differently. They use execution to refine direction instead of waiting for every detail to be settled. They act when there is enough confidence to proceed, then adjust based on what happens next.

The first move creates information the planning process couldn't produce on its own. Customer behavior gets clearer. Constraints become visible. Weak assumptions surface. What looked important in the room may matter less once the work reaches the real world, while something nobody discussed suddenly becomes the thing that deserves attention.

They shape the system as it moves.

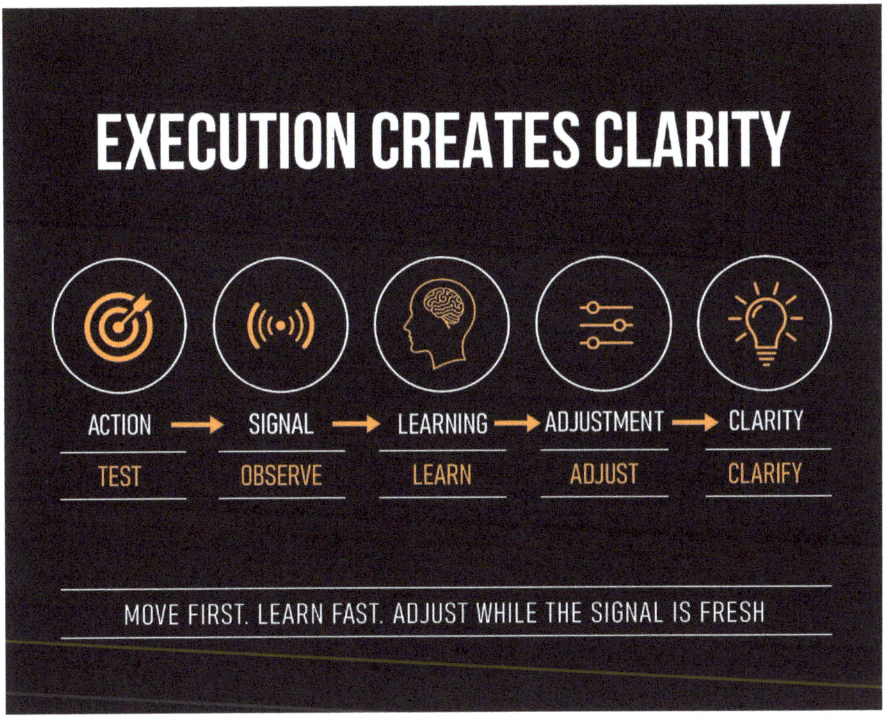

FIG 13.1 – Execution Creates Clarity. Action produces signals, learning, adjustment, and clearer direction.

Execution generates clarity. Teams move early, test quickly, and observe what actually happens. The plan isn't discarded, but it isn't protected like a museum artifact with a security guard and a velvet rope.

It changes as reality provides better information.

That doesn't mean the team is improvising wildly or treating strategy as disposable. It means the strategy is allowed to learn through contact with the market. The discipline comes from keeping the work close enough to reality that the next decision benefits from what actually happened.

Execution does more than produce output. It gives the organization a way to learn.

Every action generates feedback. Some of it appears immediately, while some takes longer to surface. The value isn't only in the action itself. It's in what the action reveals.

A small release can expose a bad assumption. A rough prototype can reveal what the polished deck missed. A conversation with a real customer can collapse three weeks of internal debate.

Reckless movement still misses the point. The goal is to act soon enough that reality can improve the work while there is still time to use what you learn.

Execution turns strategy into evidence. When action slows, feedback arrives later, and the organization loses time it could have spent correcting course.

Faster execution matters only when it improves the next decision. That is where advantage is either created or lost.

Execution Is a Feedback Engine

Organizations that treat execution as a feedback engine close the gap between intention and reality.

They design work to produce useful evidence quickly and prioritize actions that reveal information, not just actions that complete tasks. Instead of asking only what needs to be built, they also ask what needs to be learned.

That changes how teams define progress. A finished task may still have limited value if it doesn't reduce uncertainty, test an assumption, or improve the next decision. The work matters most when it gives the organization something useful to act on.

It also changes how leaders evaluate speed. The fastest team isn't necessarily the one producing the most output. It's the one turning action into insight quickly enough to adjust while the evidence is still fresh.

Feedback only creates advantage when it reaches the next decision in time to change it. If the finding gets buried in a retrospective, delayed until the next planning cycle, or separated from the people who can act, the organization has generated information without learning much from it.

Output still matters, but insight determines direction.

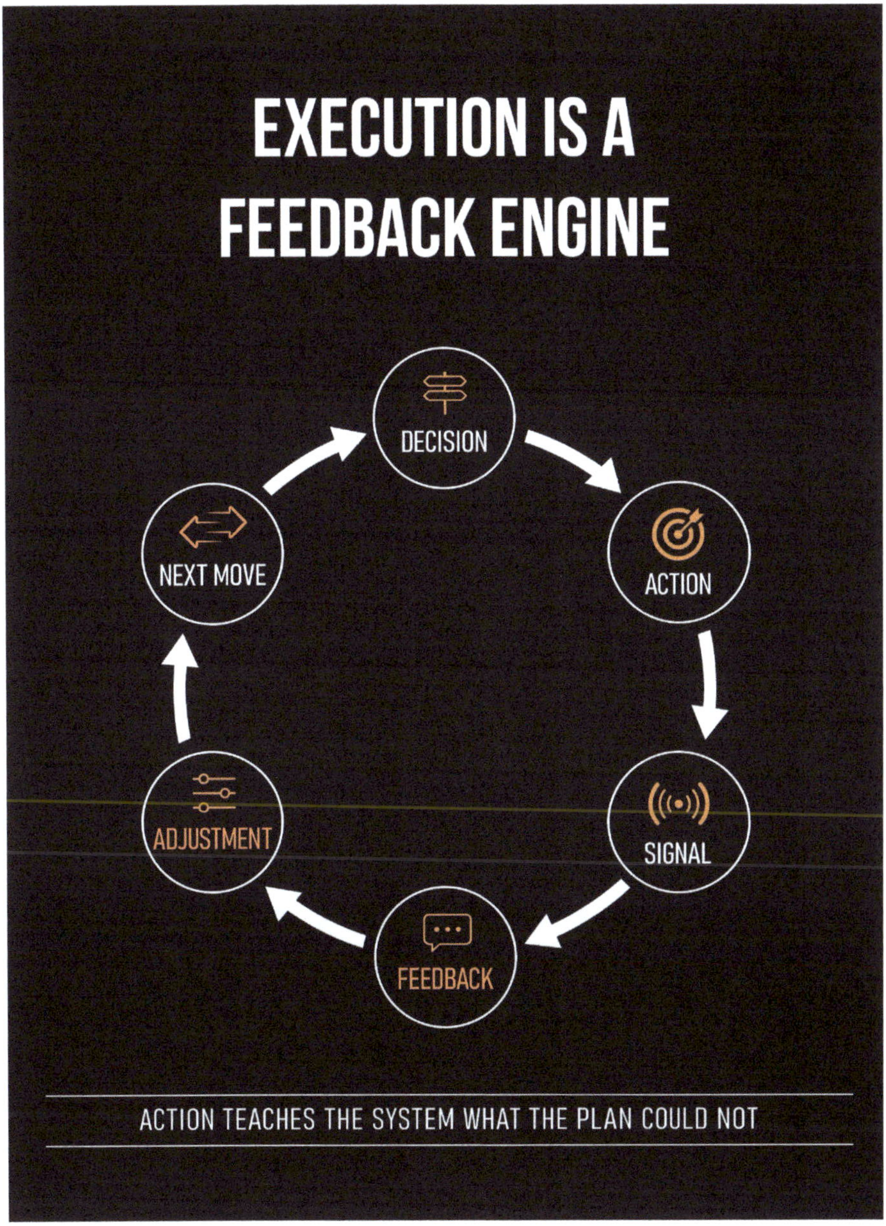
EXECUTION IS A
FEEDBACK ENGINE
DECISION
ACTION
SIGNAL
FEEDBACK
ADJUSTMENT
NEXT MOVE
ACTION TEACHES THE SYSTEM WHAT THE PLAN COULD NOT

At small intervals, differences in execution speed can be hard to see. One team moves slightly faster while another takes more time. The gap appears minor.

Over time, it compounds.

The faster team runs more cycles, tests more ideas, gathers more feedback, and adjusts more frequently. Each cycle builds on the last.

The slower team works through fewer cycles, investing more time in each move and aiming for precision. The result may be more refined in isolation.

It's less informed by reality.

This is where separation happens.

Not from one heroic decision.

From many small actions that produce better information sooner.

Execution speed determines how quickly an organization learns. Learning determines how effectively it adjusts. Adjustment determines how well it performs.

The relationship is direct.

When execution slows, everything downstream slows with it, even if the organization believes it's protecting quality.

Permission Is Not a Strategy

Execution slows for predictable reasons. Direction is unclear. Dependencies are heavy. Approval paths are long. Teams are uncertain about what "good" looks like.

Each one creates hesitation.

Each one adds time between decision and action.

That time is often treated as necessary.

Sometimes it is.

Often it's just the organization asking for permission from itself.

AI-native organizations identify friction in the path of execution and remove it where it doesn't add value. They clarify direction, reduce dependencies, and define standards that allow teams to act without waiting for validation.

Coordination still exists.

Delay doesn't get a free pass.

Hesitation usually comes from the fear of getting it wrong. That concern is real. Big decisions carry risk, and the cost of failure can be significant. In slower systems, that risk was often managed by trying to build more certainty before acting.

In faster systems, that same instinct creates delay that's harder to recover from.

A more effective approach is to break decisions into smaller moves. Act on what can be tested quickly, observe what happens, and adjust before committing further. That reduces the cost of being wrong and increases the speed of correction.

The system becomes more resilient, not because mistakes disappear, but because they are contained.

Ownership Keeps Execution Aligned

Execution also breaks down when ownership is unclear. Work moves, but accountability does not. Tasks are completed. Outputs are generated. Progress appears to be happening.

Then something doesn't align.

And no one moves first.

That's where drift starts.

AI amplifies this. Outputs are generated quickly. Tasks move forward with less direct involvement. It becomes easier for work to progress without a clear owner guiding it.

That creates risk that can be hard to detect in the moment. The work keeps moving, but the authority to correct it gets harder to find.

High-performing teams maintain ownership even as execution speeds up. Every meaningful action has a responsible party. Every outcome is tied to someone who can adjust it.

Speed doesn't remove accountability.

It increases the need for it.

AI accelerates execution by reducing the time required to move from idea to output. It allows teams to test concepts quickly and lowers the barrier to action.

What it does not do is decide what should be executed.

That still requires judgment.

Teams that use AI effectively treat it as a way to increase the number of useful cycles they can run, not as a replacement for thinking. They reduce the time between thought and action while maintaining control over direction.

Execution only creates value when the system learns from the work.

Each action should produce information. Each cycle should reduce uncertainty. Each adjustment should bring the organization closer to what actually works.

When execution is designed this way, speed becomes an advantage.

Activity volume isn't the key.

The advantage comes from turning execution into learning.

That's the difference between activity and execution. One fills the calendar. The other teaches the system what to do next.

If nothing is learned, the motion was just expensive noise.

Signals from the Edge

In a restaurant technology product company, I worked with a team that spent nearly a quarter refining a feature before release. The requirements were detailed. The edge cases had been argued into submission. We shipped it polished.

Within a week, customers were using it differently than we expected, and we were rebuilding the parts we had just perfected.

We had not avoided cleanup.

We had delayed learning until it was expensive.

Reality, as usual, had not agreed to the schedule.

The response was mixed, and the team had to adjust after roughly 80 percent of the effort had already been spent.

The team wasn't wrong about the work. The intent was responsible.

But we had spent so much time protecting the first release that we delayed the one thing that would have made the work better. Contact with actual users.

By the time feedback arrived, the team was no longer learning early.

Delay wasn't the only cost. Spending most of the budget before reality had a chance to improve the work was the real damage.

The correction came too late.

That one stayed with me because the lesson arrived only after the expensive part was already over.

That's a brutal way to learn what a smaller release could have taught sooner.

The work was polished before it was informed.

I've seen what happens when a team takes the opposite approach.

A team released a simpler version earlier. It wasn't complete, but it showed how users actually behaved, and the team adjusted in real time.

Within a few cycles, the outcome was stronger.

Both teams worked hard.

Only one used execution to learn.

Mindset Shift to Adopt

Stop treating execution as the final step.

Start using execution as the way strategy learns.

Tools to Try This Week

Break one large initiative into smaller, testable actions.

Define what "good enough to learn" looks like before starting.

Decide what signal will trigger the next adjustment.

Use AI to accelerate the creation of testable outputs, but keep human judgment responsible for what gets tested and why.

Do This Now

Pick one over-refined initiative. Choose something waiting for more polish, more alignment, or more internal confidence.

Find the smallest useful action. Identify the move that would produce real feedback from the world, not just another internal opinion.

Move it this week. Do not protect the first version like it's sacred. Put it in contact with reality.

Use the signal. Capture what the action revealed and use it to adjust the next move.

Key Takeaways

Action creates feedback, and feedback sharpens direction.

Smaller moves reduce risk and increase adaptability.

Friction between decision and action delays learning.

AI increases execution speed, but judgment still decides what deserves to move.

MACH-10 Leadership Moves

Move one initiative before it feels complete. Use execution to create evidence, not just output.

Shorten the gap between decision and action. Reduce the time between knowing enough and doing something useful.

Remove one source of friction in your execution path. Cut the dependency, approval, or handoff that slows learning without improving the result.

Define who can adjust the work after it ships. Execution needs someone close enough to correct it while the signal is still fresh.

You don't need perfect execution.
You need execution that learns.

Chapter 14

You Can't Run Every Decision

Leadership Is About How Decisions Get Made

Leadership changes once the organization starts moving faster than you can personally control.

In a slower environment, leaders could stay close to decisions. They reviewed work, approved direction, and validated outputs before anything moved forward. That created consistency because there was still enough time to operate that way.

The model depended on having that time.

That assumption doesn't hold anymore.

There are too many decisions, too much information, and too much activity for one person or leadership group to stay involved at that level of detail. Trying to do it doesn't create control. It creates delay, even when the intent is quality.

That's what makes this hard.

The instinct is to hold on, stay involved, add checkpoints, and maintain visibility by inserting yourself into the process. It feels responsible because it looks like leadership staying close to the work.

Eventually, it becomes the bottleneck with a calendar invite and a very official-looking agenda.

AI-native leadership depends less on staying close to every decision and more on shaping how those decisions happen when you aren't in the room.

That doesn't mean leaders become less involved. It means they get involved earlier, where their judgment can shape the rules, boundaries, and decision paths before the work starts moving. By the time every output reaches the leader for review, the organization has already paid the cost in waiting, handoffs, and lost momentum.

The role shifts from controlling individual actions to designing the conditions that produce them. Leaders define how work gets created, how decisions get made, and where judgment has to enter before something moves.

The test of leadership isn't how many decisions run through you. It's how many good decisions can happen without you.

That changes where attention goes. Leaders spend less time touching every piece of work and more time shaping the conditions around it.

Teams don't need constant approval. They need clear direction.

When priorities are specific enough to guide trade-offs, teams can act without waiting for oversight. They know what matters, what can flex, where judgment is required, and where the boundary sits.

That reduces intervention and exposes weak leadership faster.

Traditional leadership models rely on approvals to maintain control. Work progresses once it has been reviewed, and decisions are validated before execution. In stable environments, that can create consistency.

At speed, it creates bottlenecks.

Approvals compress decision-making into specific moments and force work to wait for validation that may not change the outcome. AI-native leaders replace many of those approvals with guardrails.

Guardrails define the boundaries within which teams can operate. They clarify what's acceptable, what isn't, and where additional scrutiny is required. Inside those boundaries, people are expected to act without waiting for permission.

Control shifts from something applied after the fact to something embedded in the work itself.

When the guardrails are clear, teams don't have to guess whether they're allowed to proceed. They know which decisions they own, which risks require escalation, and which outcomes cross the line. That cuts down the constant back-and-forth that makes leaders feel informed while leaving everyone else stuck in place.

It also makes accountability easier to see. If a team acts within the agreed boundaries, the organization can support the decision even when the outcome isn't perfect. When someone steps outside them, the problem becomes visible quickly.

That's a different kind of control. It's proactive, designed, and in place before pressure arrives.

The goal isn't to eliminate control or rely on trust alone. It's to build an environment where trust and control reinforce each other.

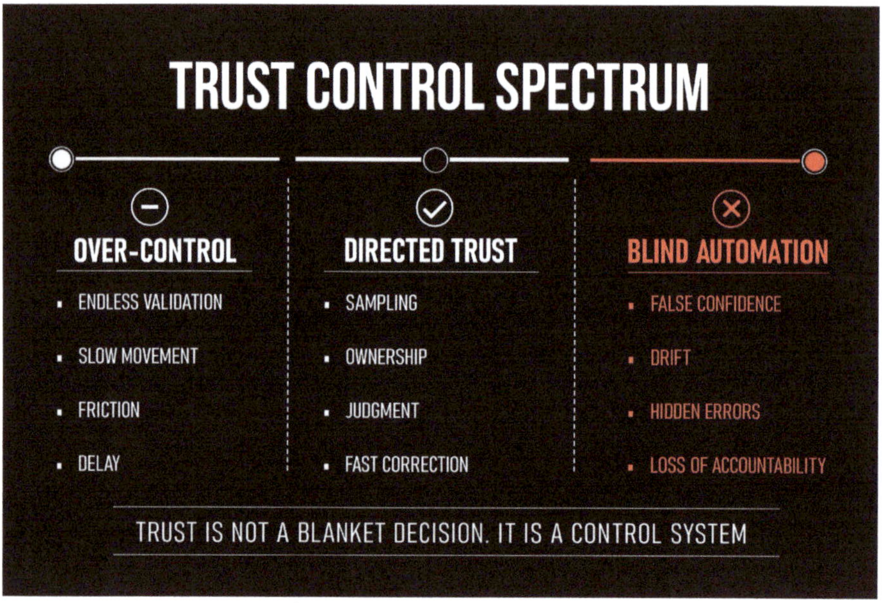

FIG 14.1 – Trust Control Spectrum. Trust works best when leaders avoid both over-control and blind automation.

Guardrails Make Accountability Visible

Guardrails don't remove accountability. They make it visible.

When the boundaries are clear, ownership becomes easier to see. Teams know what they can decide, what requires escalation, and where the line sits. The organization doesn't have to stop every time uncertainty appears like it found a suspicious bag at airport security.

Leaders still have to stay engaged, but the way they engage changes.

As operations scale, distance increases. Leaders rely more on summaries, dashboards, and synthesized outputs, which creates efficiency while also creating blind spots. The farther removed they become from the actual work, the harder it is to notice subtle changes in how things are behaving, especially when the reporting layer still looks clean.

AI-native leaders manage that distance deliberately. They don't review everything. They sample.

They stay close to the edges and look for evidence of how the work is actually behaving, not only how it's being reported. They inspect the places where outputs meet customers, decisions meet consequences, and assumptions meet reality. That gives them a chance to catch drift without turning every workflow back into an approval queue.

Leadership responsibility doesn't shrink as execution accelerates. It becomes more focused.

When something breaks, the issue is rarely limited to a single decision. The failure often reveals something deeper: responsibility may be unclear, feedback may arrive too late, or a guardrail may be missing, vague, or routinely ignored.

Those are leadership signals.

The stronger response is to address the pattern that produced the outcome. Tighten the boundary, clarify who owns what, or fix the path that allowed the problem to travel. Otherwise, the same weakness eventually returns wearing a different shirt.

Attention Must Move

Leaders still make decisions, but they can't make all of them.

Their attention has to shift toward the calls that define direction, allocate resources, set constraints, and shape how the organization operates. That's where judgment has the greatest leverage. Everything else should sit as close to the work as the consequences allow.

That takes discipline because not every decision needs escalation, not every issue deserves executive attention, and not every update needs a meeting where twelve people nod at the same slide like it just revealed fire.

Knowing where to engage is as important as knowing how.

A common failure is staying deeply involved in decisions that should happen independently while missing the handful of choices that shape everything around them. The calendar fills up, leadership feels busy, and the organization becomes increasingly dependent on people who were supposed to be designing it to operate without them.

Executive attention is expensive, even when nobody puts it on the P&L. Spend it on decisions where judgment can change the outcome, not on approving something three capable people already agreed on Tuesday. If the CEO is choosing button copy while a market shift goes unexamined, congratulations: the organization has successfully escalated the wrong problem.

AI-native leadership requires a reallocation of attention. Leaders spend less time managing execution and more time defining priorities, constraints, responsibility, and the mechanisms that bring useful feedback back into the organization.

A leader's effectiveness shows up in what happens when they aren't there. If decisions stall in their absence, too much authority is concentrated at the top. If ordinary work requires constant intervention, the boundaries aren't clear enough. If outcomes vary wildly from team to team, the operating rules probably aren't doing enough work.

Those failures appear in execution, but they usually begin in leadership design.

The strongest leaders aren't trapped inside every decision. They build an organization capable of making good ones without them, then stay close enough to recognize when the conditions change and the design needs adjustment.

Leadership now means ensuring decisions can be made effectively without you. That requires clear direction, structure, guardrails, and responsibility, along with the discipline to step back when things are working and step in when they begin to drift.

That balance defines AI-native leadership.

Signals from the Edge

In a restaurant technology scale-up environment, I watched leadership teams stay deeply involved in day-to-day decisions as AI tools and faster workflows entered the system.

Teams kept waiting for approval, work accumulated between checkpoints, and the organization became more constrained as speed increased. The leaders were trying to preserve quality, but every approval also taught the team that movement still required permission.

The healthier pattern looked different.

Leadership focused on defining direction, setting guardrails, and clarifying which decisions belonged closer to the work. Feedback loops were reinforced, ownership became easier to see, and leaders engaged selectively based on what the system was telling them instead of inserting themselves into every process.

The shift also changed what leaders could see. Instead of spending their time clearing routine work, they could focus on exceptions, weak signals, and recurring problems that revealed where the operating model still needed attention. That made leadership more useful because intervention happened where it could improve the organization, not merely keep the queue moving.

That's an underrated benefit of getting leaders out of the approval business: they regain enough altitude to notice the building is leaning. Clearing twenty routine decisions before lunch may feel productive, but it's a lousy trade if nobody notices that the same exception has appeared six times, customers are responding differently, or an entire workflow is surviving because Karen in operations has apparently become critical infrastructure.

It also forced teams to build their own judgment instead of borrowing it from leadership. When every meaningful call gets pushed upward, people get very good at preparing recommendations and very bad at making decisions. Giving them real authority inside clear boundaries changed that. They had to weigh trade-offs, live with outcomes, and correct their own misses instead of treating escalation like a corporate eject button.

The deeper change was cultural. Once people realized the guardrails were real, they stopped treating autonomy like a temporary privilege that could disappear the first time something went sideways. Teams became more willing to surface problems early, admit what they didn't know, and correct course without first building a legal defense for the original decision.

That gave them more confidence because they weren't constantly guessing whether every move would be second-guessed later. Clear boundaries let people act with speed while still knowing when to stop, escalate, or ask for help.

Over time, that created a stronger organization, not just a faster one. Leaders spent more time on the handful of issues that genuinely required their judgment, while teams became more capable of handling the rest without waiting for permission.

That allowed the organization to move faster without losing coherence. The difference came down to leadership design.

Mindset Shift to Adopt

Stop trying to control every decision.

Stop being the checkpoint.

Start designing the system that makes good decisions without you.

Tools to Try This Week

Define clear priorities that can guide decisions without requiring additional input.

Replace one approval step with a guardrail that makes the boundary clear.

Sample enough outputs to understand how the system is behaving instead of reviewing everything.

Clarify decision ownership across one team so people know who can move and when escalation is required.

Define where the team can act without waiting for you.

Do This Now

Pick one decision your team waits for you to approve. Choose something that shouldn't require your direct involvement every time.

Define the guardrail. Clarify the boundary that would allow the team to make the decision without you.

Name the owner. Assign one person accountable for the decision and the outcome.

Clarify escalation. Make clear when the decision should come back to you.

Step back and observe. Watch how the system performs without turning the test into another approval loop.

Key Takeaways

Leadership shifts from control to system design.

Clarity enables speed more than oversight.

Guardrails replace approval gates.

Sampling helps leaders stay close to reality without reviewing everything.

MACH-10 Leadership Moves

Define the boundaries for independent action. Make clear where your team can move without waiting.

Turn one approval into a guardrail. Define the boundary clearly enough that the team can move without asking again.

Reallocate your attention toward higher-impact decisions. Spend less time approving movement and more time shaping the system.

Strengthen one feedback loop that informs direction. Make sure signals reach the people who can act on them.

Measure how the system performs without you. If it stalls when you step back, the system needs better design.

You do not lead every decision.
You build the system that makes better decisions possible.

Chapter 15

The Coordination Tax

Why Speed Gets Slower as Organizations Grow

Scale rarely kills speed through one dramatic decision. It usually does it politely.

A handoff appears here, a dependency gets added there, and one more stakeholder "just needs visibility." Before long, the work isn't moving through the organization. It's asking for permission from every floor of the building.

That's the trap.

What works inside a small team rarely survives unchanged inside a larger one. Small teams run on shared context. People hear the same conversations, see the same trade-offs, and understand what matters without needing every decision translated into process.

As the organization grows, that shared understanding fragments. More people enter the picture, decisions cross functions, and dependencies multiply. The work still advances, but now it passes through more rooms, more checkpoints, and more stretches of invisible waiting.

The organization looks larger, more professional, and more mature.

It also gets slower in the places that matter most.

Scale doesn't create all of that friction by itself. It spreads the cost across enough people and steps that no one person can see the whole thing.

In a small team, a stalled decision is usually obvious. The blocker is in the room, on the call, or sitting three seats away eating pretzels while everyone waits for an answer.

At larger scale, the slowdown gets harder to trace. It hides inside handoffs, fuzzy responsibility, dependency chains, and calls that somehow require five people when one used to be enough. Work advances, pauses, and waits, but no single delay feels large enough to explain why everything takes longer.

The organization simply feels heavier.

Everyone just knows the decision left the building.

That's usually when companies make the problem worse. They add control, process, checkpoints, and review steps to create consistency. For a while, that can stabilize things.

It also adds latency everywhere.

Eventually, the latency becomes the constraint. Execution accelerates, output rises, and expectations climb, but coordination doesn't improve at the same rate.

That's when the gap opens.

Some parts of the organization move quickly while others wait for context, approval, or a dependency to clear. The faster the execution layer gets, the more obvious the coordination burden becomes.

The dangerous part is that every delay still looks reasonable on its own. One review adds a day. One dependency adds a meeting. One handoff creates another chance for context to thin out. Everyone feels only the individual steps while the work absorbs the total cost.

Velocity dies when coordination outpaces execution.

This is the Coordination Tax™.

It doesn't appear neatly on a dashboard, and you won't find it as a line item in the budget. It shows up as time: the extra meeting that didn't change the decision, the dependency that never needed to exist, or the gap between knowing what to do and actually doing it.

Every dependency adds overhead. Every additional stakeholder adds processing time. Every review step stretches the distance between decision and action.

You can see the tax by counting approval layers, handoffs, and the waiting time between them. At small scale, the organization may be able to absorb that cost. In larger environments, it compounds until coordination starts consuming more energy than the work itself.

Congratulations.

The meeting has become the product.

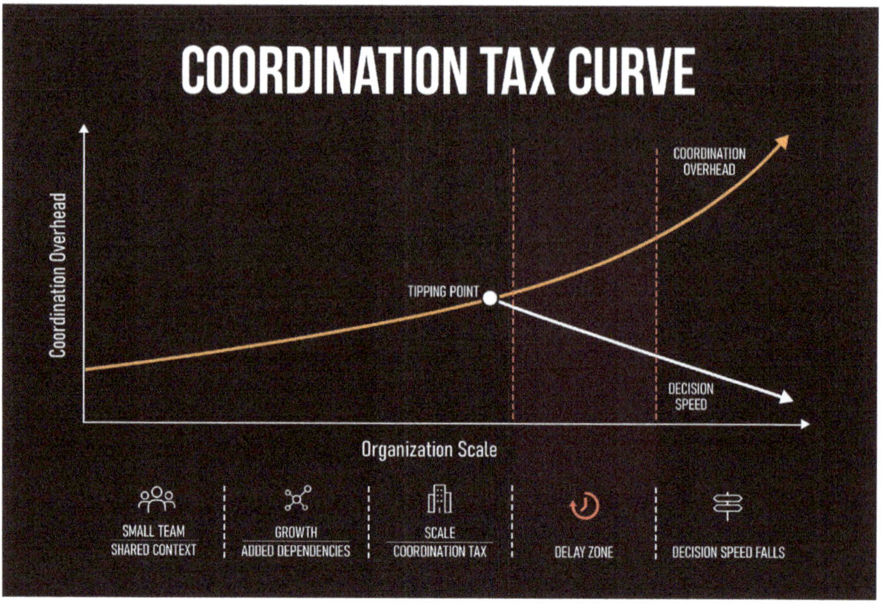

FIG 15.1 – Coordination Tax Curve. Coordination overhead rises as organizations scale, while decision speed falls when coordination exceeds the value it creates.

Large systems still need coordination. Waste begins when that coordination consumes time without changing the outcome.

Leaders have to decide where consensus genuinely matters, where it only slows things down, which decisions need shared context, and which ones need a clear owner and a call.

Those questions are uncomfortable because coordination often feels safer than ownership. It spreads responsibility, brings more people into the room, and gives everyone a chance to weigh in. Sometimes that improves the decision.

Other times, it just gives delay better attendance.

The problem is that participation can keep expanding long after the answer has stopped improving. Another stakeholder adds context, another review adds caution, and another meeting gives everyone one more chance to say the same thing with slightly different nouns. The process looks thorough, but the decision isn't getting better.

Leaders have to know when coordination has finished its job. Once the relevant context is in the room, the risks are understood, and the owner is clear, more participation starts creating drag. At that point, the meeting is no longer supporting the decision. It's circling the airport waiting for a runway that has been open the entire time.

Scale also changes the impact of mistakes.

In a small team, errors are usually contained. You catch them early, the blast radius stays limited, and the correction can happen without much disruption.

At greater scale, the same mistake travels farther. It crosses teams, affects more customers, enters more workflows, and becomes harder to unwind. The cost is no longer limited to the original error. It grows with the distance the mistake travels before anyone catches it.

Speed still matters, but risk becomes harder to ignore. Organizations often respond by slowing everything down and adding approval layers across the board.

That may reduce some risk.

It also creates the risk of being late.

In an environment that's already moving, lateness carries its own cost. The leadership challenge is to manage speed and control together without allowing either one to become the default response to every decision.

At small scale, shared context does much of the work. People hear the same conversations, see the same information, and often reach similar conclusions because they are operating from a common view of the business.

As that shared view fragments, alignment gets harder.

Different teams see different parts of the organization. Information gets filtered along the way, assumptions drift, and what feels obvious in one corner may be invisible in another. Decisions that should reinforce each other start to diverge because people are working from different versions of reality.

AI can surface more information, but it can't create shared understanding on its own.

Leaders who handle this well don't force alignment everywhere. They reinforce the signals that matter, clarify which decisions should remain consistent across the organization, and define where teams have room to operate independently.

The difference between a startup and an enterprise goes beyond size. It also includes tolerance for risk.

Startups can often move quickly because the cost of being wrong is more contained. They can test, adjust, and iterate without creating consequences that ripple across thousands of customers, regulators, partners, and employees. Enterprises play on a larger field. Brand exposure, compliance, operational complexity, and customer impact all raise the price of a bad call.

That requires structure, but the structure can't become the reason everything stops moving.

The principles still hold. Speed matters. Ownership matters. Feedback matters. Judgment matters. What changes is where each one gets applied.

That is where leadership shows up: knowing where speed creates advantage, where coordination earns its keep, and where control is necessary because the cost of being wrong is too high.

Scaling an AI-native organization requires more than repeating what worked early. Leaders have to redesign how the operating model behaves under complexity. They need to decide which decisions stay centralized, which move closer to the work, where validation genuinely protects the business, and where it is simply another layer of ceremonial armor.

Those are structural choices, and they determine whether speed survives scale.

AI doesn't solve any of this.

It makes the pattern visible faster.

If the organization is designed for flow, that flow can scale. If it is built around unnecessary coordination, the friction will scale faster than the advantage.

Signals from the Edge

I've seen this pattern in early-stage AI companies and larger enterprise environments. One startup hit it as the team grew from a small founding group into multiple functions.

Dependencies appeared. Decisions that used to happen in a conversation started requiring meetings. A product call that once took ten minutes now needed input from engineering, customer success, sales, and operations. What used to be solved through shared context now needed coordination.

Speed did not disappear.

It got stuck.

The team still had urgency, but the system had no clean way to use it.

The team tried to fix it with process, and that worked for a while. Things stabilized, but the system lost the responsiveness that created the initial advantage.

I've seen the opposite, too. An enterprise didn't try to force startup speed across the board. They focused on where speed actually mattered, clarified ownership, and reduced coordination where it didn't add value.

The system moved faster than it had before, and it held.

In another case, automation scaled across multiple teams almost overnight. Output increased, dashboards improved, and throughput went up. Within weeks, a different pattern showed up. Teams spent more time reconciling inconsistencies than building new work. Decisions required more coordination, not less.

Nothing looked obviously broken, but everything was slower where it mattered.

The automation had accelerated production without simplifying the system around it. More work entered the organization, but ownership, standards, and decision paths hadn't been redesigned to handle the volume. The result was faster output feeding a slower coordination layer.

The teams eventually improved the system by reducing unnecessary dependencies, assigning clearer owners, and defining where consistency mattered enough to justify review. They also stopped treating every difference as a defect. Some variation was harmless, while other variation created real customer, operational, or strategic risk.

That distinction mattered. Without it, the organization was holding a committee meeting every time two dashboards wore different socks. Once leaders clarified which inconsistencies required action, teams could correct the important ones without turning every edge case into a cross-functional summit.

Mindset Shift to Adopt

Stop assuming what worked early will scale.

Start designing for what breaks when it does.

Tools to Try This Week

Map one decision path.

Mark where the work waits after the answer is already clear.

Remove one dependency that adds delay without improving the decision.

Clarify which decisions need alignment and which need a single owner.

Adjust one workflow to match the reality of scale, not the fantasy of simplicity.

Do This Now

Pick one decision path that feels slower than it should.

Count every handoff, dependency, and approval step.

Identify one coordination step that doesn't change the outcome.

Assign a single owner for the next decision in that path.

Key Takeaways

Scale introduces coordination cost, and that cost compounds.

Friction becomes distributed and harder to see.

Context fragments across teams.

Structure must evolve if speed is going to survive.

MACH-10 Leadership Moves

Measure coordination value. Identify where coordination is costing more than it improves the decision.

Remove one dependency layer. Cut one handoff, approval, or waiting point in a critical workflow.

Separate speed from control. Define where movement is essential and where additional review is actually required.

Strengthen shared signals. Give teams clearer inputs so they don't need constant check-ins to stay coordinated.

Redesign one scaling process. Adjust one workflow so it can operate at greater scale without adding unnecessary delay.

Small scale creates momentum.
Scale decides whether you keep it.

Chapter 16

Judgment Becomes the Advantage

AI Accelerates Everything Except What Matters Most

When AI can do the work, your judgment is the only moat left.

At this point, it's easy to overcorrect. You've seen how quickly organizations can operate, how much leverage AI creates, and how execution, strategy, structure, and scale all change when speed becomes the constraint.

The natural response is to push further through more automation, more acceleration, and greater reliance on tools that can operate faster than people.

That response is incomplete because the faster everything runs, the more the human edge matters.

AI can process information, generate output, and identify patterns across massive datasets faster than any individual or team. That's where its strength lies. What it can't do is decide what actually matters.

That requires judgment.

Research on judgment under uncertainty points to the same reality. Better tools don't eliminate uncertainty. They change where leaders encounter it.

Judgment goes beyond analysis. It draws on experience, taste, responsibility, timing, and the ability to recognize when something that looks correct doesn't fit reality. It also allows leaders to act when information is incomplete and the outcome remains uncertain.

That capability doesn't scale the way computation does. It has to be practiced.

This is where differentiation starts to shift. As AI becomes more capable, leaders can mistake cleaner output for better thinking. The work gets faster, the options look stronger, and the recommendations sound more complete. The smoother the answer feels, the easier it becomes to confuse acceptance with evaluation.

Once everyone has access to the same intelligence layer, output stops being the advantage.

Judgment becomes what separates one organization from another.

Two teams can begin with the same AI-generated market analysis and end up in completely different places.

The first accepts the recommendation because it's clear, structured, and supported by enough data to feel safe. They act quickly, but they stay inside the assumptions they were handed.

The second uses the same analysis differently. They ask what the model missed, test the timing, and look for the constraint hidden behind the clean answer. They ask whether the recommendation still holds if the market shifts, the customer behaves differently, or the competitor gets there first.

The input is the same. The interpretation isn't, and neither is the outcome.

That's where the real difference appears. It doesn't come from access to the better-looking answer. It comes from knowing what to do with it.

The Human Edge Directs the System

The human role is to direct AI, not compete with it.

That means knowing where to trust the output and where to question it, when to act quickly and when to slow down, and how to interpret what the technology is telling you against what is actually happening.

That is where control comes from: speed directed by judgment.

Not every decision can be reduced to logic. Some involve trade-offs that don't appear cleanly in the data. Others require nuance, timing, customer perception, and an understanding of what will hold up in practice rather than what merely looks correct in isolation.

That is where taste enters.

Taste is the ability to see beyond the available options. It is the judgment to look at two paths that both make sense and still recognize which one is more likely to survive contact with reality.

AI can generate the options, but it can't assign the weight.

That remains the decision.

This is the real Matrix problem for leaders. It isn't about dodging bullets. It's about recognizing when the simulation looks clean and still asking whether it maps to reality.

Cleaner Output Isn't Stronger Judgment

The technology produces and people react. That can work longer than it should, until it breaks, usually all at once.

When an organization shifts from shaping decisions to responding to outputs, depth starts to erode. People stop interrogating the work. They stop asking which assumptions support the recommendation, what information is missing, or what AI may be smoothing over because it doesn't understand the stakes.

The output gets cleaner, but the thinking gets thinner.

That's the failure mode.

If judgment has been outsourced, there's nothing left to catch the gap between what looks right and what's right.

The organization may operate quickly. It may look impressive. It may even come with a beautiful dashboard, because apparently every bad idea deserves a chart now.

More Average, But Faster

Speed without judgment still has no idea where it's going.

Leaders who ignore this lose differentiation. The output remains efficient. It remains technically correct. It also starts to look like everything else.

AI operates on patterns. It identifies what is most likely based on what it has seen. That works when the patterns hold, and it breaks when they don't.

When everyone uses the same tools, leans on the same patterns, and accepts the same kinds of answers, the work starts to converge. It becomes smoother, faster, and more average.

The danger isn't that the work looks bad. The danger is that it looks good enough to stop questioning.

That's how differentiation disappears.

Quietly.

The machine keeps producing, and the market stops caring.

Turns out "more average, but faster" isn't much of a moat.

Shifts in behavior, new conditions, and exceptions expose the limits of pattern recognition. That's where humans need to engage, not by rejecting the output, but by interpreting it. They have to ask whether the pattern actually applies, what might be missing, and whether the machine is confidently wrong.

Responsibility Does Not Move Into the Model

There is another layer that gets missed.

AI doesn't carry responsibility. It produces output.

The consequences belong to the people and organizations that act on it. When AI informs a decision, responsibility can start to feel distributed. The model suggested it. The data supported it. The recommendation looked sound.

But someone chose to act.

That doesn't change.

Leaders who understand this don't treat AI as cover. They use it as input while remaining responsible for what happens next. They understand that the cost of being wrong isn't absorbed by the model.

It's absorbed by the organization.

Responsibility stays human, but the ability to carry it has to be maintained.

The Judgment Gap

The gap opens when leaders still own the outcome but stop practicing the judgment required to evaluate the work.

AI can make the analysis cleaner, the options stronger, and the recommendation easier to approve. That's useful. It's also where leaders can start confusing polish with confidence.

The danger isn't that AI producces bad work.

Bad work is usually easier to catch.

The danger is work clean enough for leaders to stop pressing on it.

That's where human judgment has to remain active.

Judgment Atrophies When It Is Not Used

Human judgment also weakens when it isn't used.

As AI takes on more execution, people get fewer opportunities to practice it. Decisions happen faster, but leaders engage with them less deeply. Over time, that creates atrophy.

The technology gets better, the people get less capable, and the organization gains speed while losing the muscle that knows where to aim it.

That is the quiet danger. Judgment doesn't disappear all at once. It gets exercised a little less until the organization stops noticing what it no longer knows how to evaluate.

That decline is avoidable, but it requires intention.

Leaders need to stay engaged in the parts of the work where judgment still matters. They need to create room for thinking, not just execution. They have to challenge recommendations, question assumptions, and remain close enough to understand what good actually looks like.

That doesn't require slowing everything down. It requires knowing where depth matters.

One way to preserve that depth is to make leaders explain the call, not just approve it. What assumption mattered most? Which trade-off carried the most weight? What would have changed the decision? Those questions keep judgment active because they force people into the reasoning beneath the recommendation instead of allowing them to approve whatever landed on top.

Otherwise, executive judgment can slowly turn into the corporate equivalent of a gym membership you keep paying for because technically you still belong. The capability is there, everyone feels better knowing it exists, and nobody notices how little it has been used until somebody actually needs to lift something heavy.

Teams also need opportunities to compare what AI recommended with what actually happened. That's where judgment improves. Leaders can see which factors mattered, which assumptions failed, and where intuition added value or introduced bias. Without that feedback, approval becomes routine and judgment never develops.

The danger isn't that AI gets better at execution. It's that leaders get worse at recognizing when execution still requires interpretation.

That is how judgment becomes ceremonial.

People continue approving the work, but they stop truly evaluating it. Everything keeps moving, and the judgment muscle gets softer each time.

The most dangerous part is that this rarely feels like decline while it's happening.

It feels like efficiency.

The team gets faster answers. Leaders spend less time wrestling with ambiguity. Meetings come with cleaner summaries and better-structured options. The work feels more organized, and in many ways, it is.

But fewer people are doing the difficult interpretive work behind it.

They are reviewing instead of reasoning, approving instead of weighing, and reacting to polished options instead of forming judgment through contact with messy reality.

Judgment isn't preserved by proximity to decisions. It is preserved through use. Leaders don't stay capable because important choices pass in front of them. They stay capable because they keep practicing the uncomfortable work of deciding what matters when the answer is incomplete.

AI can make that work easier to avoid.

The risk isn't only that the machine becomes more powerful. It's that the leader becomes too removed.

The process may still ask for approval, but approval isn't the same as judgment. Approval can happen at the surface. Judgment has to reach into the assumptions, trade-offs, timing, consequences, and pieces of reality the model can't feel.

When leaders stop doing that, they don't lose judgment all at once. Then one day, AI produces something clean, confident, and wrong, and no one left in the loop has the muscle to challenge it.

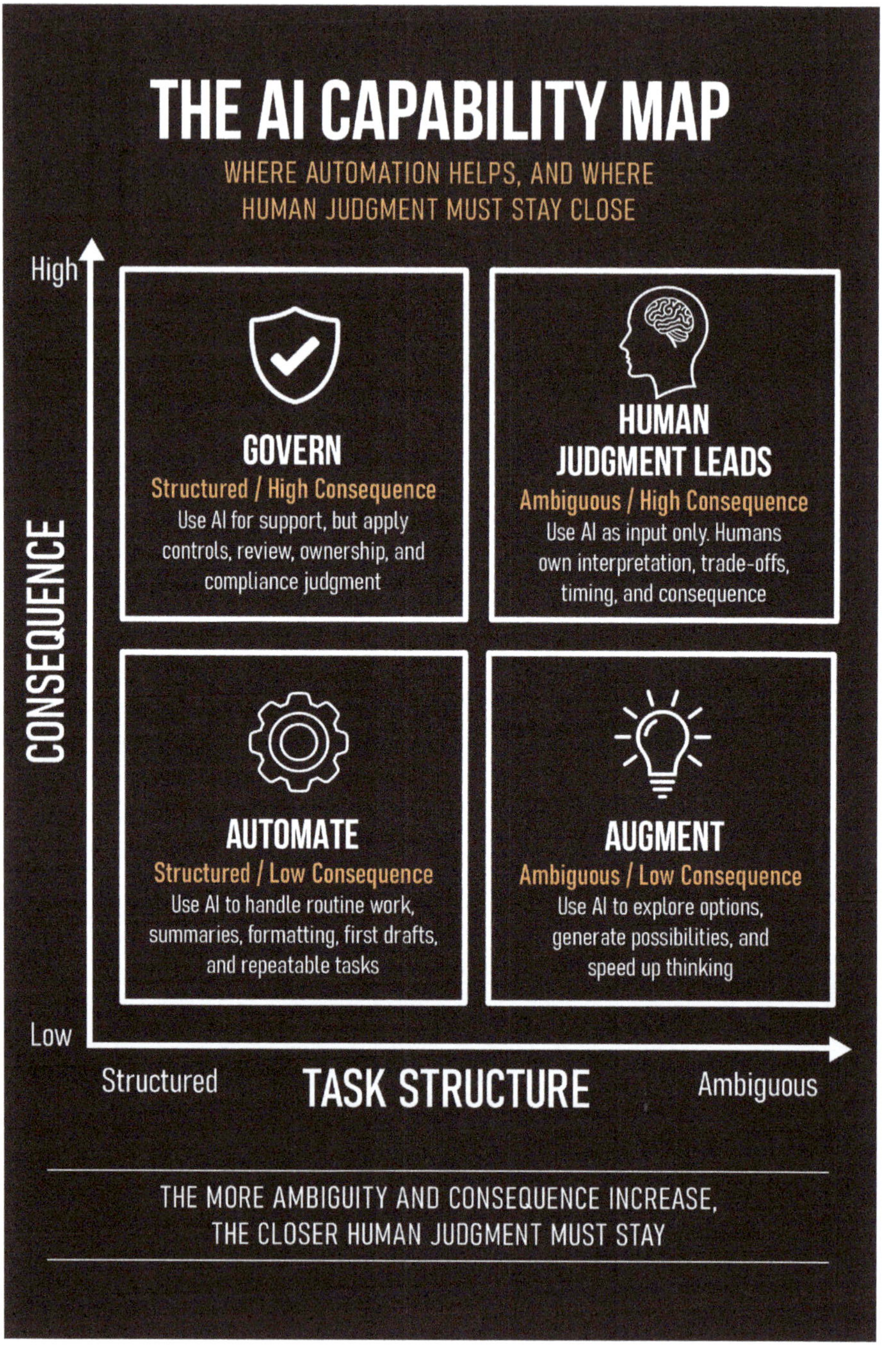
THE AI CAPABILITY MAP
WHERE AUTOMATION HELPS, AND WHERE
HUMAN JUDGMENT MUST STAY CLOSE
High
GOVERN
Structured / High Consequence
Use AI for support, but apply controls, review, ownership, and compliance judgment
HUMAN JUDGMENT LEADS
Ambiguous / High Consequence
Use AI as input only. Humans own interpretation, trade-offs, timing, and consequence
CONSEQUENCE
AUTOMATE
Structured / Low Consequence
Use AI to handle routine work, summaries, formatting, first drafts, and repeatable tasks
AUGMENT
Ambiguous / Low Consequence
Use AI to explore options, generate possibilities, and speed up thinking
Low
Structured
TASK STRUCTURE
Ambiguous
THE MORE AMBIGUITY AND CONSEQUENCE INCREASE,
THE CLOSER HUMAN JUDGMENT MUST STAY

AI is strongest in structured, low-consequence tasks. Human judgment matters most where ambiguity and consequence are high.

The value created by judgment doesn't show up in a single decision.

It compounds.

Teams that rely only on speed and output can look competitive in the short term. They produce more. They move quickly. They look efficient.

Judgment is what produces outcomes that survive contact with reality.

Signals from the Edge

Across product, pricing, go-to-market, and AI automation work, I've seen teams use the same tools, the same data, and the same market signals and still make very different decisions. The pattern showed up most clearly in decisions involving pricing, positioning, customer trade-offs, and AI-generated recommendations that looked complete before anyone had tested the assumptions underneath them.

The roadmap looked coherent. The sequencing made sense. The internal logic held together. But one assumption beneath it had never been tested against real customer behavior.

That was where judgment should have entered.

The output looked complete enough to move. The recommendation was clean. The answer felt good enough to accept.

The better question was which assumption would change the decision.

What did the model miss about the customer, timing, market, or consequence?

What would make this dangerous if it were wrong?

The same input produced different outcomes because one team accepted the answer and the other practiced judgment against it.

Mindset Shift to Adopt

Stop treating AI output as the differentiator.

Start treating judgment as the advantage that determines whether the output matters.

Tools to Try This Week

Take one AI-generated recommendation your team is considering.

Ask which assumption underneath it would create the most damage if it were wrong.

Identify what the system can't know from the prompt, the data, or the pattern alone.

Compare the recommendation against timing, context, consequence, and customer reality.

Ask what a leader would still need to decide even if the recommendation were technically correct.

Do This Now

Pick one decision supported by AI. Choose a decision where the output looks clean enough to accept.

Find the missing context. Identify what the system may not understand about the customer, market, timing, or internal constraint.

Name the judgment call. Write down the part of the decision that can't be delegated to the system.

Move with judgment. Use the output, but don't hide behind it.

Key Takeaways

When everyone has access to the same intelligence layer, output stops being the advantage.

Judgment determines what matters, what changes, and what carries consequence.

Clean answers can weaken interpretation when leaders stop practicing it.

The leader's role is to decide what the output means.

MACH-10 Leadership Moves

Interrogate one clean answer. Find the assumption that would create damage if it were wrong.

Separate output from judgment. Identify what the system produced and what the leader still has to decide.

Test the context. Ask whether the recommendation survives contact with customer reality, market timing, and consequence.

Name the human judgment point. Decide where interpretation enters before the recommendation becomes action.

Protect the human edge. Use AI to increase capacity without letting judgment go soft.

The system can move faster than ever.
It still needs you to decide what matters.

Chapter 17

The Workflow Is the Transformation

Start Where Friction Is Highest

At some point, this stops being conceptual.

You can understand the shift. You can agree that speed matters. You can see why decision-making, structure, and execution need to change. None of that matters until it shows up in how work actually moves.

This is where most organizations stall, usually because they hesitate rather than disagree.

They look for a plan. A rollout. A clean sequence that takes them from where they are to where they want to be. Preferably with a steering committee and a slide called "Phase 2: Enablement," because apparently that's how adulthood happens in corporate America.

They expect this to behave like other transformations, where you define the approach, align the team, and execute against it.

That instinct creates the problem.

A rollout changes availability. Improvement changes behavior.

Workflow change creates reality.

Real improvement starts where the friction is highest. It starts where work slows, decisions sit, and everyone quietly pretends the process is normal.

This doesn't behave like a traditional transformation. There is no clean starting point or ceremonial moment where the organization flips from one model to another. What actually happens is uneven. Some teams adapt quickly. Others lag. Certain workflows improve while adjacent ones keep operating exactly as they did before.

If you wait for it to feel organized, you won't start.

The better move is to find one workflow where the cost is already visible and change how that workflow behaves. It doesn't solve everything, but it proves the organization can operate differently.

Once one real workflow starts making decisions faster, learning sooner, and wasting less motion, the organization has something more useful than a plan.

It has evidence.

That evidence changes the conversation. Teams no longer have to imagine what a faster operating model might look like because they can see one working inside the business. The debate shifts from whether the approach can work to where it should be applied next.

The teams that get through this usually do something simpler: they pick a place and begin.

A real workflow. Something important enough to be worth changing and contained enough that a bad adjustment won't set the building on fire. They go where work is slow, decisions sit, and the pain is obvious to anyone paying attention.

That's where the machinery reveals itself.

When you look closely, the problem is usually mechanical. Time disappears between steps. Information arrives and waits. A decision is needed and sits. Action gets agreed on and somehow acquires a three-day layover. Feedback exists somewhere, but it isn't connected tightly enough to the next call.

None of it looks dramatic on its own.

No one writes a postmortem about a three-day handoff. They just slowly die inside during the next status meeting.

The easiest place to start is usually not the biggest workflow. It's the one everyone already complains about, works around, or explains away. That's where the hidden operating model is easiest to see. The delays have names. The handoffs have owners. The meetings have calendar invites. The problem isn't theoretical anymore. It has a room number, a recurring agenda, and at least one person saying, "That's just how we do it."

That sentence is practically an invitation.

You don't need a new model to see the problem.

You need to look at how the work actually flows.

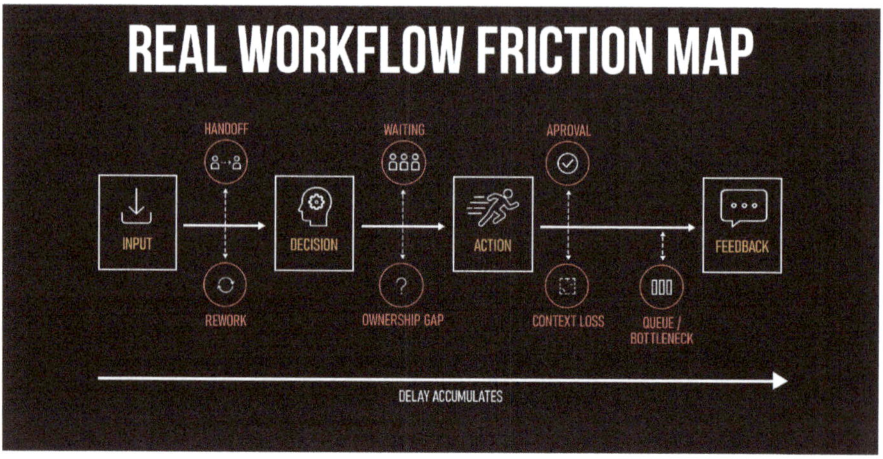

FIG 17.1 – Real Workflow Friction Map. Delay accumulates between steps.

Once you see the friction, the changes become practical.

You shorten the gaps, clarify who decides so decisions don't sit, remove steps that remain only because they've always been there, and reduce the distance between decision and action. You connect feedback directly to the next move instead of letting it wait somewhere until it becomes stale.

None of this requires a new system. It requires removing what's slowing the current one down.

This is where most teams get it wrong.

They reach for tools first.

AI becomes the starting point instead of the lever. It gets dropped into workflows that are already unclear, slow, and fragmented. The output improves, but the workflow still can't move.

That can look like progress. Usually, it's faster output moving through the same bottlenecks.

I've seen teams celebrate adoption while the actual work stayed stuck. The rollout created more activity, but it didn't create a cleaner path from decision to action. More people were using the tool, more drafts were being generated, and more summaries were being produced. The dashboard looked healthy.

The same decisions still waited for approval. The same handoffs still created delay. The same ownership gaps remained.

The tool had changed.

The workflow hadn't.

AI shouldn't be the first move. Clarity should.

Where is the process slow? Where is ownership unclear? Where is work waiting for no good reason?

Fix that first. Then apply AI where it can actually improve the path from information to action.

Otherwise, you end up automating the organizational equivalent of standing in line faster. Everyone feels more advanced because the queue has a chatbot now, but the door at the front is still locked and Gary from finance still has the key.

The second shift is making decision-making real as a working mechanism, not just a concept.

Who owns the call? What input actually matters? When is there enough information to act? What needs to move quickly, and what requires more control?

Most teams assume those answers are already clear. Usually, they aren't.

What exists instead is habit. Decisions happen a certain way because they always have. People stay involved because they've always been involved. Steps remain because no one has removed them.

When decision-making becomes explicit, hesitation drops and the path forward gets easier to see.

It also makes weak points easier to diagnose. If the evidence is clear but the decision still stalls, ownership is probably the issue. If the owner is clear but the team still hesitates, the threshold for action may be vague. Once those gaps are visible, leaders can fix the actual constraint instead of wrapping another layer of process around it.

That's why clarity has to be visible. It can't remain trapped in a leader's head, buried in a meeting summary, or reduced to the vague sense that everyone probably knows what comes next. Teams move faster when they can see what matters, understand the context, weigh the options, know the threshold, identify the owner, and act.

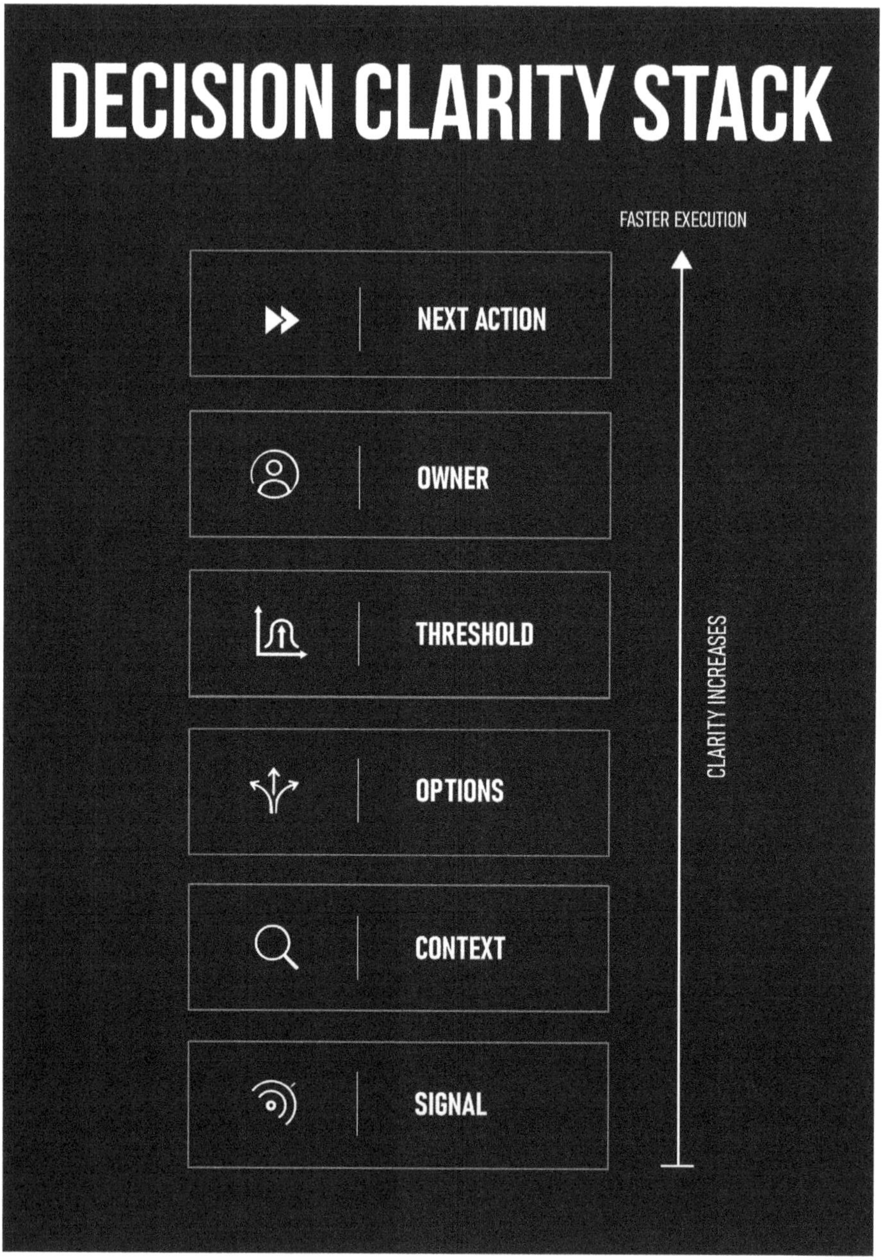

FIG 17.2 – Decision Clarity Stack. Faster execution depends on clarity moving from signal and context to options, threshold, ownership, and next action.

The third shift is where the system starts to compound.

In most organizations, feedback is delayed. Work is completed, results are reviewed later, and adjustments are made in the next cycle. That delay separates action from learning, and in a faster system, that separation gets expensive.

You have to see what's happening while it's happening.

You have to connect outcomes to decisions immediately.

You have to reduce the distance between action and adjustment.

You stop looking at the system after the fact and start observing it in motion.

None of this feels like a transformation while you're doing it.

It feels like cleaning things up.

You make decisions clearer. You remove steps that slow work down. You connect actions to outcomes more directly. You introduce AI where it reduces time instead of adding noise.

A series of small changes starts to compound.

This is where most teams make the next mistake.

They try to scale it too quickly.

They fix one workflow and assume the rest will follow. Then they push the same changes everywhere and expect consistency before the system has earned it. That creates resistance because the context is different. Dependencies are different. Risks are different.

People are different.

You don't scale the workflow.

You scale the thinking.

Where is the friction? Where is the delay? Where is the system slowing itself down?

Those questions transfer even when the exact answers don't.

Over time, something shifts.

Decisions happen faster without feeling rushed. Work moves without waiting for unnecessary alignment. Feedback shows up early enough to matter. AI is used where it creates leverage instead of noise.

It doesn't feel like a new system.

It feels like the system finally moving the way it should.

Signals from the Edge

I've been in the middle of a product rollout that looked healthy from the outside and messy the second you got close to it.

The launch moved.

The workflow did not.

Support was finding issues before Product saw them. Product was waiting on context from the field. Operations was trying to explain exceptions after customers had already felt them. Everyone was working hard, but the same problems kept bouncing between teams without a clear owner.

The release had shipped, but the system around it had not caught up.

I've also seen the better version.

A team started with one ugly post-launch handoff. Customer issues were moving from support to product to operations and back again. They mapped the mess, cut one loop, named the owner, and used AI only where it shortened the path to the next decision.

Nobody called it transformation.

The workflow just started moving.

Real change begins when one painful part of the system stops wasting everyone's time.

Mindset Shift to Adopt

Stop waiting for a complete plan.

Start improving how work actually moves.

Transformation starts when one real workflow stops pretending the friction is normal.

Tools to Try This Week

Pick one workflow and map how it really operates by separating effort time from wait time.

Look for where decisions are sitting and why.

Remove one step that no longer adds value.

Introduce AI in one place where it shortens the gap between steps.

The goal is to help the workflow move with less friction, not simply automate it.

Do This Now

Pick one slow workflow. Choose something that feels heavier than it should.

Map the path. Trace it from input to decision to action to feedback.

Find the biggest gap. Identify where time disappears between steps.

Remove one low-value step. Cut or time-box something that doesn't change the outcome.

Assign one owner. Make one person accountable for the next decision in that workflow.

Workflow Friction Scorecard

Before you call an automation successful, ask these questions.

- Does this reduce decision time, or just manual keystrokes?
- Where does the correction tax show up, and who owns it?
- Does the workflow move faster, or does the same bottleneck now receive cleaner output?
- If this breaks at scale, who knows first?

Key Takeaways

Real change starts in real workflows.

Delay lives between steps, not just inside them.

AI should reduce friction in the workflow, not decorate the bottleneck.

MACH-10 Leadership Moves

Pick one place where work is slow and fix it. Start where the system is already showing you the problem.

Name the next movable step. Make the workflow show exactly where action should happen next.

Remove one unnecessary step this week. Cut the step that adds time but not quality.

You don't need to change everything.
You need to start where it matters.

Chapter 18

Competition Now Moves at Decision Speed

Why Learning Speed Now Drives Results

The rules have already changed in practice.

You can see it in the way some teams move while others are still explaining why they're almost ready. Some organizations are learning in real time while others are aligning around plans that were current three weeks ago.

This is what a real competitive shift looks like. It rarely arrives all at once. It shows up in smaller ways: one team releases faster, a competitor adjusts sooner, or a product improves between cycles instead of between versions.

Over time, they create separation.

Most organizations misread what is happening because they focus on what they can see: features, pricing, positioning, and roadmaps. They assume competition is happening at that level, so they respond by optimizing those surfaces.

That is the visible game.

The real advantage sits elsewhere.

Competition is increasingly defined by how quickly an organization can adapt what it builds, what it learns, and what it does next.

A strong starting position still matters. A well-defined plan still matters. Scale, brand, distribution, and resources still matter. None of them, however, can protect an organization that can't adjust when the market changes.

The teams creating separation are not always the ones with the best first idea. They are the ones that can test sooner, see what actually happens, and adjust before everyone else has fully reacted. Their advantage comes from turning contact with the market into better decisions while the information still matters.

They create clarity through contact with the market.

Strategy research describes part of this as dynamic capability: the ability to sense what is changing, seize the opportunity, and reconfigure before competitors do. MACH-10 leadership brings that logic down into the decision system, where strategy stops being an abstract capability and starts showing up in the speed and quality of everyday calls.

Annoying for the teams still waiting on the perfect deck, but here we are.

That is the new competitive edge.

Knowledge is only part of the equation. The organization still has to turn what it learns into action before everyone else learns the same thing.

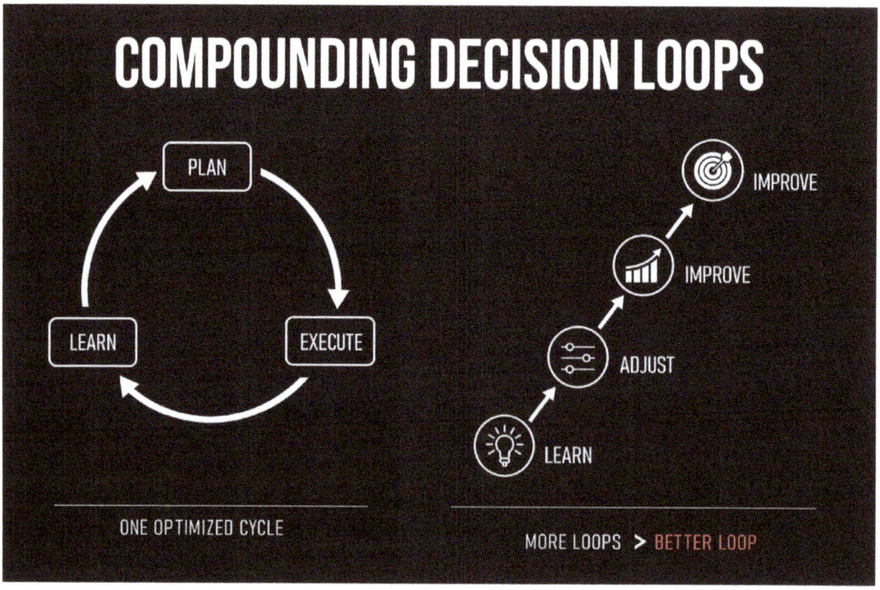

FIG 18.1 – Compounding Decision Loops Model Revisited. Advantage compounds when faster decisions create faster learning loops.

This separation doesn't appear all at once. It compounds.

Each cycle builds on the last, and each adjustment improves the next move. Over time, the distance between teams that operate this way and teams that don't becomes increasingly difficult to close.

The first few missed loops may look harmless.

Eventually, they start shaping the market.

Then the gap gets teeth.

By the time the difference becomes obvious, the faster organization has already learned through several cycles you haven't run.

The competitor isn't beating you because its first answer was better. It's beating you because it reached its fifth answer while you were still defending your first.

That's what most organizations underestimate. They assume they can catch up. They see a competitor moving faster and believe enough effort, budget, or executive attention will close the distance.

Often, they're already solving the wrong problem.

On the surface, they're competing against a product. In reality, they're competing against an organization that learns faster than they do.

You can copy a feature, but you can't easily copy the operating rhythm that keeps improving it.

Screenshots are easy. Operating rhythm is harder.

That is why speed compounds into position. The first release may not be better. The second gets closer. By the third, the team understands the customer more precisely. By the fourth or fifth cycle, the product is being shaped by evidence its slower competitor simply doesn't have yet.

At that point, the competitor is no longer responding to your product.

You are responding to its pace.

Startups Feel It First

This shift doesn't affect every organization in the same way.

Startups usually feel it first. They operate closer to the edge, with fewer layers and less coordination overhead. When they adopt AI-native ways of working, the impact can be immediate. Decisions happen sooner, feedback arrives faster, and a small team can gain ground on much larger competitors before those competitors fully recognize what is happening.

Enterprises experience the same pressure under very different conditions.

They have scale, resources, customer relationships, distribution, and brand trust. They also have complexity. Coordination takes longer, decisions cross more boundaries, risk is harder to isolate, and the cost of being wrong can travel much farther.

That doesn't make speed impossible. It makes selective speed more important.

Enterprises do not win by pretending to be startups in hoodies making decisions from beanbag chairs. They create separation by accelerating the places where timing actually changes the outcome.

That means identifying which decisions can happen closer to the work, where faster learning improves competitive position, and where additional control earns its cost. It also means using AI to shorten the distance between insight and action without filling the organization with so much automated noise that somebody eventually needs another dashboard to explain the dashboards.

The organizations that get this right don't try to become something they're not.

They adapt where it counts.

The Market Gets Louder

That distinction matters because the competitive landscape is getting more crowded.

Barriers to entry are falling. AI lowers the cost of building, shortens the path from idea to product, and allows smaller teams to produce work that once required much larger organizations.

That means more competitors, more experiments, and more chances for someone else to discover the thing you are still discussing internally.

In that environment, standing still isn't neutral. It costs position.

The dashboard may look fine, the roadmap may be advancing, and the teams may be busy. Everyone may have even updated their TPS reports, cover sheet and all.

The market remains unmoved by this achievement.

This is where strategy, execution, and leadership stop being separate conversations. Strategy still defines direction, but it has to remain connected to current conditions. Execution still produces outcomes, but it also needs to generate useful evidence. Leadership still shapes how the organization operates, but now it has to do so at a pace the environment will actually tolerate.

When those pieces drift apart, capability starts losing value.

Adaptation speed determines whether the organization can still use what it knows before the knowledge gets old.

Adaptation Is No Longer Optional

You don't get to decide whether the environment accelerates. You decide how the organization responds.

Some companies will keep operating the way they always have, planning, aligning, and executing through cycles that feel familiar. For a while, they'll look stable. Then the distance will start to show.

Others will swing too far in the opposite direction. They'll chase speed without enough structure, act before understanding the consequences, and produce more without getting meaningfully better. That approach has a shelf life too.

The strongest position belongs to organizations that can accelerate without becoming reckless and adapt without losing their center. That takes more than faster tools or shorter cycles. Leaders have to know which calls can happen quickly, which deserve more scrutiny, and how new evidence gets pulled back into the next decision before the window closes.

That discipline matters even more when conditions change again. The organization doesn't have to rebuild its operating model from scratch or launch another transformation program with a logo, a kickoff deck, and twelve months of enthusiastic status updates. It already knows how to sense change, make the call, act, and correct while the opportunity is still alive.

It also changes how uncertainty feels inside the business. A new condition doesn't automatically trigger three more weeks of planning or a mad sprint in whatever direction looked exciting Tuesday afternoon. There is enough structure to keep operating while leaders test what matters, contain the downside, and adjust before uncertainty turns into delay.

The organizations that define what comes next will behave differently. They'll apply speed where timing creates advantage, use judgment where the consequences are highest, and build the capability to learn faster than slower competitors can respond.

The ability to move, learn, and adjust continuously without losing coherence is the competition now.

Signals from the Edge

I was at a social media startup that entered a market dominated by larger competitors with more resources, broader distribution, and greater reach.

Our advantage was faster learning.

We ran shorter cycles, released more often, and adjusted constantly. At first, the product wasn't always better, but the operating rhythm was.

Over time, the product caught up and then moved past competitors that looked stronger on paper.

That experience stayed with me because the larger players weren't beaten by one feature. They were beaten by a system that learned faster than theirs.

The difference showed up in how quickly each organization could turn evidence into another move. We didn't need to be right on the first pass. We needed to learn before the larger system finished debating whether the signal was strong enough to deserve attention.

That rhythm also created confidence of a different kind. It didn't come from believing every decision would be correct. It came from knowing the team could detect what wasn't working, adjust quickly, and keep the cost of being wrong from compounding. I've also seen large organizations recognize the shift early and focus on where speed could actually improve their position.

They didn't try to change everything at once. They pushed decision-making closer to the teams operating at the edge, shortened feedback loops, and used AI to accelerate learning instead of only increasing execution.

The strongest enterprise teams treated that speed as a design problem rather than a cultural slogan. They clarified where authority could move outward, where risk still required central judgment, and how signals would return quickly enough to shape the next decision.

The enterprise didn't need to imitate startup speed. It needed to move faster than its old structure allowed.

Mindset Shift to Adopt

Stop competing only on what you build.

Start competing on how quickly your system adapts.

Tools to Try This Week

Identify one area where your team is adapting slower than the market.

Shorten the time between action and feedback.

Compare your cycle time to a real competitor, not to your internal planning calendar.

Remove one delay that isn't improving the decision.

Use AI to accelerate one part of your learning loop, but measure whether that speed turns into better adjustment.

Do This Now

Pick one faster-moving competitor. Choose a competitor that's moving faster than expected.

Find where their system learns faster. Look past the feature and identify where their cycle appears shorter than yours.

Choose one adaptation-critical workflow. Pick a place where speed of learning actually affects competitive position.

Remove one delay between signal and action. Cut or time-box the step that slows movement without improving the decision.

Measure the next cycle. Track whether the next move improves faster because the loop is shorter.

Key Takeaways

Competition is increasingly defined by adaptation speed.

Small differences in speed compound into structural advantage.

Organizations compete against learning systems, not just products.

Advantage comes from continuous movement, learning, and adjustment.

MACH-10 Leadership Moves

Identify where speed creates competitive advantage. Not every workflow matters equally. Find the one where faster learning changes position.

Shorten one competitive response loop. Reduce the time between market signal, decision, action, and adjustment.

Give market-facing teams one decision they can make without escalation. Let the people closest to the signal act while it's still fresh.

Align execution with real-time feedback. Do not let learning wait for the next planning cycle.

Treat learning speed as a competitive metric. Measure how quickly the system converts signal into adjustment.

You aren't competing in the same environment anymore.
Act like it.

Chapter 19
The Leadership Mandate

What This Actually Requires From You

At this point, none of this should feel abstract anymore.

You know the organization has to move. You know decisions are the constraint. You know speed without judgment breaks things, and judgment without speed loses relevance.

The real question is whether you're willing to change how you lead.

That's the uncomfortable part. Most leaders love the idea of faster decision-making until the organization asks them to stop being the speed bump. They want teams to move faster, but they also want the same visibility, approval points, and familiar rituals that made them feel in control.

That's where the mandate becomes real.

The organization won't behave differently if every meaningful call still has to pass through the same old gates, get its badge scanned, and wait for leadership to finish lunch.

The mandate is simple now.

Stop treating speed as pressure. Start treating it as the environment.

Most teams don't fail this shift because they disagree with it. They fail because they try to bolt it onto the way they already operate. The approval paths stay in place. The old habits survive. Alignment gets inserted into every meaningful move, and leaders talk about speed while protecting the machinery that keeps slowing everything down.

The mandate isn't to talk faster.

It's to build an organization that knows when it can act without waiting for permission.

ALIGNMENT IS NOT PROGRESS

That's where leadership actually shows up.

It doesn't show up in the AI tools you announce, the language you use at the town hall, or the slide with a rocket ship on it, although someone will absolutely make that slide.

Leadership shows up in what you remove, what you stop doing, and how you change the conditions under which decisions get made. If those conditions don't change, neither will the organization.

The first mandate is clarity.

What actually matters? What deserves speed? What deserves scrutiny? What can happen without you, and what needs a clear escalation path?

If you don't draw those lines, teams will try to accelerate everywhere and create noise instead of progress. Output rises, activity looks like momentum, and everyone gets very busy. Then somebody eventually asks why the business still feels stuck.

Clarity prevents that. It doesn't come from more information or another dashboard. It comes from defining what matters enough to act on, what requires judgment, and what deserves to be ignored.

If your team can't answer those questions without asking you, the problem isn't communication.

It's dependency.

Clear direction also gives people something useful when reality stops matching the plan. They don't have to wait for another instruction every time conditions change because they understand the priorities, boundaries, and trade-offs well enough to make a bounded call. That's what turns leadership intent into action instead of another document people remember to open three days after it would have helped.

The second mandate is deciding where you belong.

Most leaders say they want teams to move faster, then insert themselves into the exact decisions creating the delay. They ask to be looped in, want visibility, and need to understand before anything happens. The instinct is understandable.

It's also expensive.

I've seen this with leaders who were genuinely trying to help. They cared about quality, context, and making sure the right call was made. They were also carrying old habits, and without realizing it, they had become part of the slowdown.

Every "loop me in" became a pause. Every "run it by me first" became a queue. Every "let's make sure we're aligned" created another place for momentum to leak.

The leader became the checkpoint the organization had to survive before it could act.

The cost went beyond delay. It taught dependency. Teams learned that authority still lived somewhere above them, so even capable people began waiting for validation they didn't actually need.

You can't ask for speed while requiring yourself in the loop before speed is allowed to happen. You can't tell teams to act faster while designing an organization where decisions only move when your calendar opens up.

At some point, the choice gets uncomfortable. You either remain the system, or you build one that works without you.

That's the difference between being the star player and becoming the coach. At some point, your job is no longer to take every shot. It's to create better shots without needing the ball in your hands on every possession.

Stepping back can feel like losing control. In reality, it changes where control lives.

Instead of controlling decisions after they're made, you define how good decisions should happen. You establish the boundaries, make the trade-offs visible, and explain what "good enough" means so teams aren't waiting for permission every time the situation changes.

That's what creates speed.

Less hesitation. Less waiting. Better judgment at the point where the work actually happens.

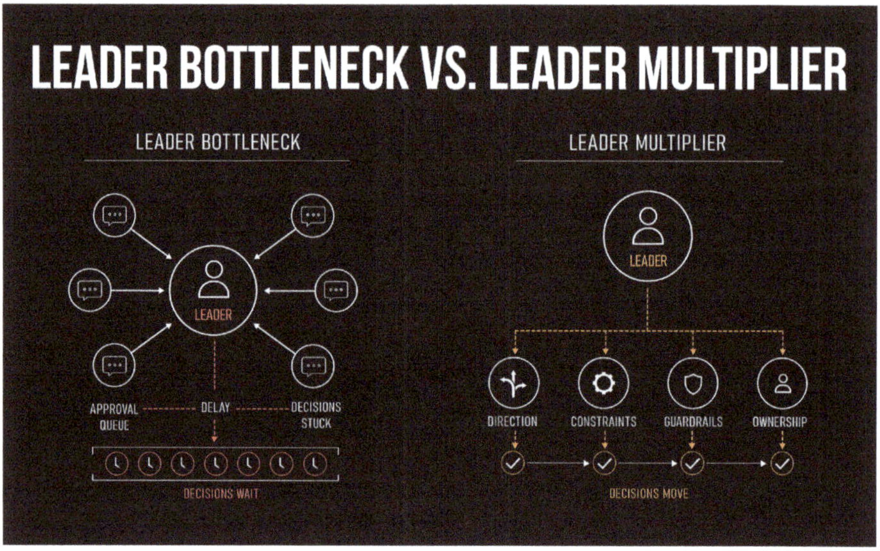

FIG 19.1 – Leader Bottleneck vs. Leader Multiplier. Decisions wait when leaders sit at the center. They move when leaders define direction, ownership, and feedback.

In one system, everything waits for the leader. In the other, the leader has shaped an organization that can move.

That doesn't mean leaders disappear. It means they stop acting like a human toll booth for every decision with a pulse.

Accountability is where this either holds or breaks.

As AI becomes more embedded in how work gets done, ownership becomes easier to blur. The output looks clean, the recommendation feels complete, and the decision appears obvious.

AI doesn't own the decision. Your team does.

If something goes wrong, the model can't absorb the impact. The business does. The customer does. The people responsible for the outcome do.

If responsibility gets fuzzy, the organization can adapt quickly in the wrong direction. By the time anyone notices, the mistake may already have momentum.

Clear ownership prevents that. It tells the team who can act, who has to challenge the output, and who is responsible for correcting the call when reality pushes back. Without that clarity, speed turns into drift and everyone discovers accountability at the exact moment something breaks.

The third mandate is continuous adjustment.

You have to be willing to change how the organization operates while it's running. This isn't a one-time transformation. It's ongoing operating work.

What works now won't hold forever. New tools will appear, new constraints will surface, and new failure modes will emerge. The way your team operates today probably won't be the best way six months from now.

Treat this like something you fix once, and you'll fall behind again.

Some teams are attached to how things worked before. Some leaders are comfortable being involved in everything. Parts of the organization were rewarded under the old model, and those habits won't disappear because somebody renamed the initiative and gave it a logo.

Resistance doesn't automatically mean you're wrong. Sometimes it means you've reached the part of the organization that has been protected for too long.

Ignoring that resistance won't help. Working through it is part of the mandate.

There will also be moments when things feel less stable. Decisions will happen faster. Some will be wrong. Some changes won't work as expected, and some teams will get ahead of the rest of the organization.

That pressure is where leaders are most tempted to retreat into the old model. More approvals. More checkpoints. More people copied "for visibility." Suddenly the organization is rebuilding the exact machinery it was trying to escape, only now everyone gets to call it risk management.

That is the moment to redesign instead of retreat.

Leadership keeps mistakes from compounding.

That requires staying close enough to see where decisions slow, where responsibility gets fuzzy, where AI creates real leverage, and where it simply creates more noise. Pay attention to where teams hesitate because they don't understand the boundary and where the same issue keeps returning under a different name.

Then adjust how the organization operates.

Not the speech. Not the slogan. The actual machinery.

That is what the mandate comes down to.

You are not here to manage a slightly faster version of the past. You are here to build an organization capable of operating at the speed of what is happening now.

That requires clarity, discipline, and the willingness to release control where it creates delay while applying it where the consequences demand it.

The work is easy to name and hard to practice because it is uncomfortable.

Unfortunately, that is where most real leadership lives.

Signals from the Edge

I worked with a company that wanted faster execution but still had too many decisions flowing through leadership by default. The team wasn't slow because people lacked urgency. They were slow because the decision path required permission even when the work was already clear enough to move.

We focused on one decision category that kept creating delay. Routine product and customer-facing workflow decisions that didn't need executive review every time. I advised the team to replace approval with direction, constraints, and escalation rules. The question changed from "Who needs to sign off?" to "What boundary would let the team move without waiting?"

That shift mattered quickly. Decision ownership moved closer to the work. Leaders stayed involved where consequence was high, but they stopped sitting in the middle of decisions that were better handled by the team closest to the signal.

Within a few cycles, the workflow was moving materially faster. Time to decision dropped by roughly 40%, and the team was able to move from signal to action without waiting for the next leadership checkpoint.

The faster cycle also exposed problems sooner, which meant the team could correct them before they turned into another round of executive review.

That was the lesson.

Speed did not come from telling people to move faster.

It came from redesigning the conditions that had been teaching them to wait.

Mindset Shift to Adopt

Stop trying to be involved in every important decision.

Start designing the conditions that allow good decisions to happen without you.

Tools to Try This Week

Identify one decision that currently requires your approval and remove yourself from it.

Define what "good enough" looks like for a decision your team is waiting on.

Clarify ownership where it currently feels shared.

Set the boundary between what needs your judgment and what only needs your awareness.

Remove one step that exists only to create visibility for you.

Do This Now

Pick one decision path. Choose a decision path that currently runs through you.

Define direction and constraints. Give the team what they need to act without guessing.

Name the owner. Make one person accountable for the decision and the outcome.

Clarify escalation. Tell the team when the decision should come back to you.

Step out and observe. Leave the decision path and watch how the system performs.

Key Takeaways

Speed is created by conditions, not intent.

Ownership must remain clear as systems accelerate.

Control belongs in the decision system, not every individual decision.

MACH-10 Leadership Moves

Remove yourself from one decision path this week. Replace your involvement with clear direction and constraints.

Reinforce ownership across your team. Make sure every meaningful decision has someone accountable for the outcome.

Identify one place where your involvement is slowing movement. Decide whether the team needs your judgment, your awareness, or neither.

You're either accelerating the system.
Or teaching it to wait.

Chapter 20

The MACH-10 Leadership Model

Where Speed, Judgment, and Responsibility Meet

By now, this shouldn't feel like a collection of ideas. It should feel like a different way to see how work actually moves.

Speed changes the clock. Judgment determines where human force still matters. Responsibility keeps movement from turning into noise.

That is the structure behind the MACH-10 Leadership Model. It shows where speed creates momentum, where judgment has to enter, and where responsibility turns motion into results.

Push one too far or neglect another and the whole thing starts to wobble.

The Model Only Works When All Three Move Together

The point of the model isn't to optimize any one dimension in isolation.

Speed without judgment creates motion without aim. Judgment without speed produces insight that arrives too late. Responsibility without either one leaves someone accountable for a decision that never really gets anywhere.

Speed creates movement. Judgment gives it direction. Responsibility makes sure somebody owns what happens next.

That combination matters because organizations rarely fail from having none of the three. They fail when one becomes dominant. Speed takes over and the team starts breaking things faster. Judgment takes over and every decision turns into a graduate seminar. Responsibility gets emphasized without enough authority or pace, and someone ends up owning a beautifully documented delay.

When all three stay connected, decisions move with clarity, consequence, and force.

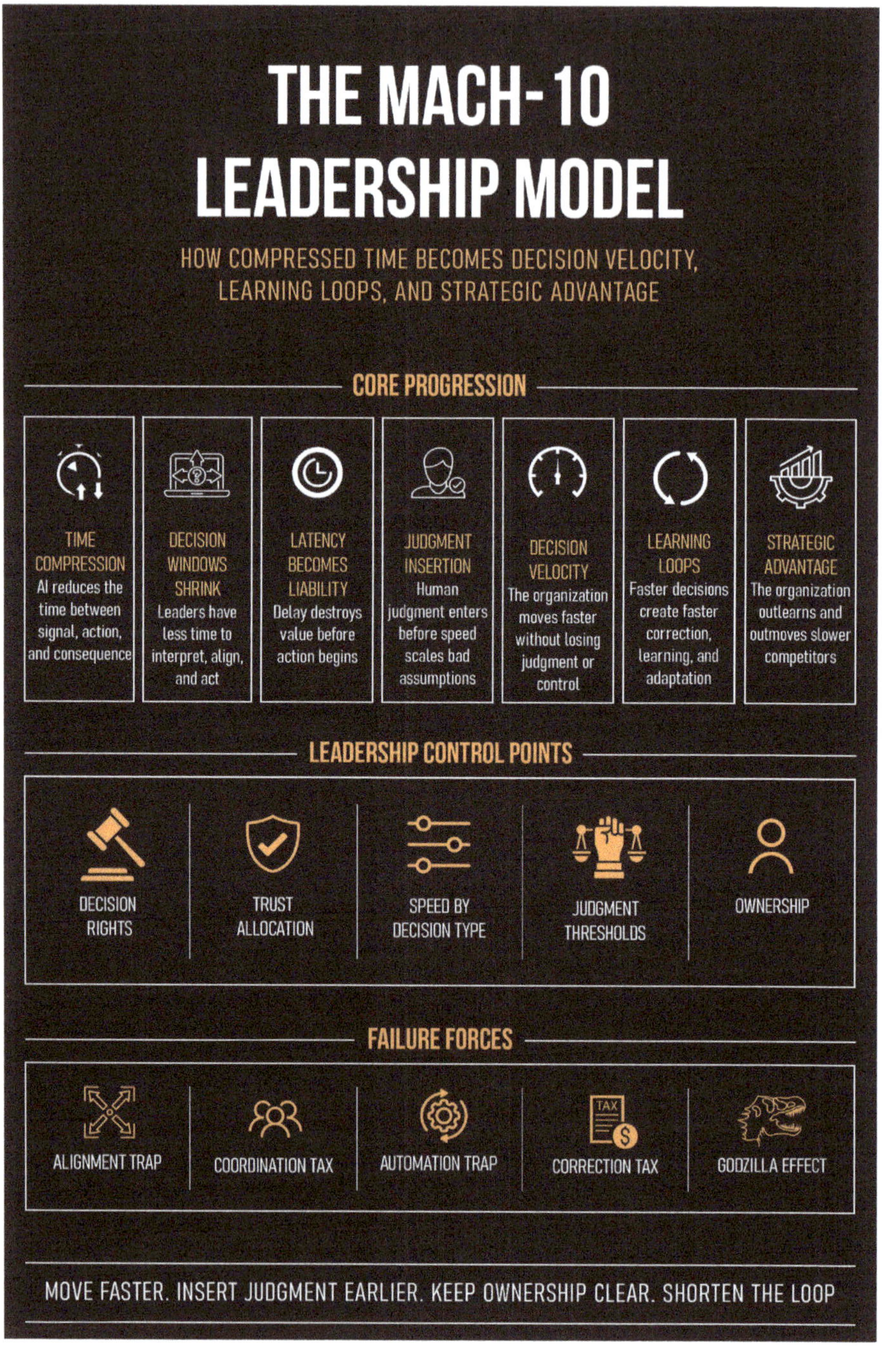
THE MACH-10
LEADERSHIP MODEL
HOW COMPRESSED TIME BECOMES DECISION VELOCITY,
LEARNING LOOPS, AND STRATEGIC ADVANTAGE
CORE PROGRESSION
TIME COMPRESSION
AI reduces the time between signal, action, and consequence
DECISION WINDOWS SHRINK
Leaders have less time to interpret, align, and act
LATENCY BECOMES LIABILITY
Delay destroys value before action begins
JUDGMENT INSERTION
Human judgment enters before speed scales bad assumptions
DECISION VELOCITY
The organization moves faster without losing judgment or control
LEARNING LOOPS
Faster decisions create faster correction, learning, and adaptation
STRATEGIC ADVANTAGE
The organization outlearns and outmoves slower competitors
LEADERSHIP CONTROL POINTS
DECISION RIGHTS
TRUST ALLOCATION
SPEED BY DECISION TYPE
JUDGMENT THRESHOLDS
OWNERSHIP
FAILURE FORCES
ALIGNMENT TRAP
COORDINATION TAX
AUTOMATION TRAP
TAX
CORRECTION TAX
GODZILLA EFFECT
MOVE FASTER. INSERT JUDGMENT EARLIER. KEEP OWNERSHIP CLEAR. SHORTEN THE LOOP

There is a version of this where nothing really changes.

You adopt a few tools, talk about speed, adjust a workflow or two, and declare progress before the organization itself has changed. On the surface, it looks responsible. Underneath, the same operating model remains intact.

Decisions still take too long. Ownership stays unclear. Work continues waiting in places where it should move.

That version is easy.

It is also how organizations fall behind while congratulating themselves for modernizing. The language changes, but the operating physics remain the same.

The better path is more direct.

Start removing what slows the work down. Clarify what matters. Define how decisions get made so they don't sit unresolved. Act before everything feels fully complete, then use what happens next to improve the next call.

Do that consistently, and the organization begins behaving differently because the conditions around the work have changed.

That is where speed actually appears. It becomes something the operating model produces rather than a slogan leaders repeat.

If the organization cannot move, it doesn't matter how strong the strategy looks, how good the ideas are, or how capable the team is. A decision that arrives after the moment has passed is no longer a strategic asset.

It is paperwork with confidence.

Speed is what allows everything else to matter.

That is also where judgment enters.

AI will produce outputs that look complete. It will reduce effort and make it easier to act without stopping to question what is happening. That leverage is real, and so is the risk.

Judgment keeps the work grounded. It helps leaders notice when something looks right on the surface but doesn't fit the reality around it. It allows them to act when information is incomplete and prevents the organization from drifting toward whatever is easiest to produce.

Judgment doesn't scale like computation, which is why it has to be applied deliberately.

Responsibility sits beneath both.

Decisions happen faster, outputs travel farther, and mistakes compound more quickly. The impact of each action expands with the reach of the organization.

That impact doesn't belong to the tool. It belongs to the people who choose to act.

You don't get to hand responsibility to a model, a process, a dashboard, or something that appears to be working simply because nobody has complained yet.

Most teams lose this balance in predictable ways. Some push for speed and lose control. Work accelerates, but the pieces stop holding together. Decisions happen, but they don't reinforce one another. The organization becomes active without becoming more effective.

Others overcorrect. They add steps, broaden alignment, increase review, and slow everything down to avoid mistakes. The process becomes safer on paper while losing the timing required to matter.

Some teams distribute responsibility so widely that ownership disappears. Everyone is informed, everyone is included, and everyone can explain the process.

No one owns the outcome.

That is how a decision goes to committee and comes back as fog wearing a name badge.

The teams that figure this out do something simpler.

They act once there is enough clarity. They question outputs that feel too easy, assign ownership before momentum diffuses, and correct quickly when something doesn't hold. Over time, that becomes normal operating behavior rather than something the organization has to summon during a crisis.

That consistency creates separation.

The advantage doesn't come from moving faster in every situation. It comes from reducing the number of decisions that depend on extraordinary effort, executive intervention, or a last-minute push to get across the line. Routine work travels with less resistance, while the decisions that genuinely need judgment still receive it.

That changes the role of leadership. Leaders spend less time forcing ordinary calls through the organization and more time shaping the conditions that allow good ones to happen without them. They clarify boundaries, make ownership visible, and step in where context or consequence demands it instead of attaching themselves to every decision by default.

It also changes what teams expect from themselves. Clarity, ownership, and timely decisions stop feeling like exceptional behavior reserved for high-pressure moments. They become part of the operating standard, which makes delay, ambiguity, and ceremonial alignment much easier to spot.

The model has to keep running because this is something you maintain, not something you install once and admire.

The environment will keep changing. New tools will appear, new constraints will emerge, and new failure modes will surface. What works now won't hold forever. Treat this as a one-time change and the organization will gradually return to what it already knows how to do.

That is the annoying thing about operating models.

They have muscle memory.

Old approval paths reappear. Old meeting habits return. Old ambiguity finds a new outfit and starts calling itself alignment again.

Once the new standard takes hold, though, regression becomes easier to see. A decision that sits too long, an owner who remains unclear, or a review step that adds no value no longer feels normal. It becomes evidence that the organization is drifting and needs correction before the old habits settle back in.

That is why the model has to be practiced rather than admired.

There will be pressure in both directions. One side will tell you to slow down, wait for more information, align more broadly, and make sure everything is understood before acting.

Some of that is necessary.

Much of it is habit wearing a responsible outfit.

The other side will push you to accelerate everything, automate too quickly, trust outputs because they look clean, and skip steps that still matter in the name of speed.

That creates a different kind of failure.

The work is knowing the difference, and that distinction only becomes visible in real decisions with real consequences.

You don't need another framework. You need to apply the ones already in front of you.

You don't need more alignment by default. You need clarity, ownership, and action.

This is what the model comes down to: acting in time, applying judgment where the consequences demand it, and keeping responsibility visible through what happens next.

Every organization eventually reaches the same moment. The evidence is visible, the information is incomplete, and the trade-offs are understood well enough to act.

The organization can keep the loop open, or it can close the decision and go.

That is where MACH-10 leadership shows up.

It requires enough speed to matter, enough judgment to stay grounded, and enough ownership to carry the consequence.

The point isn't to move faster for its own sake. The point is to build an organization where speed, judgment, and responsibility remain connected when pressure rises. That is what keeps motion from becoming noise and leadership from becoming ceremony.

The Leadership Triangle

Leadership at MACH-10 means balancing three forces that determine whether speed creates progress or instability:

Speed.

Judgment.

Responsibility.

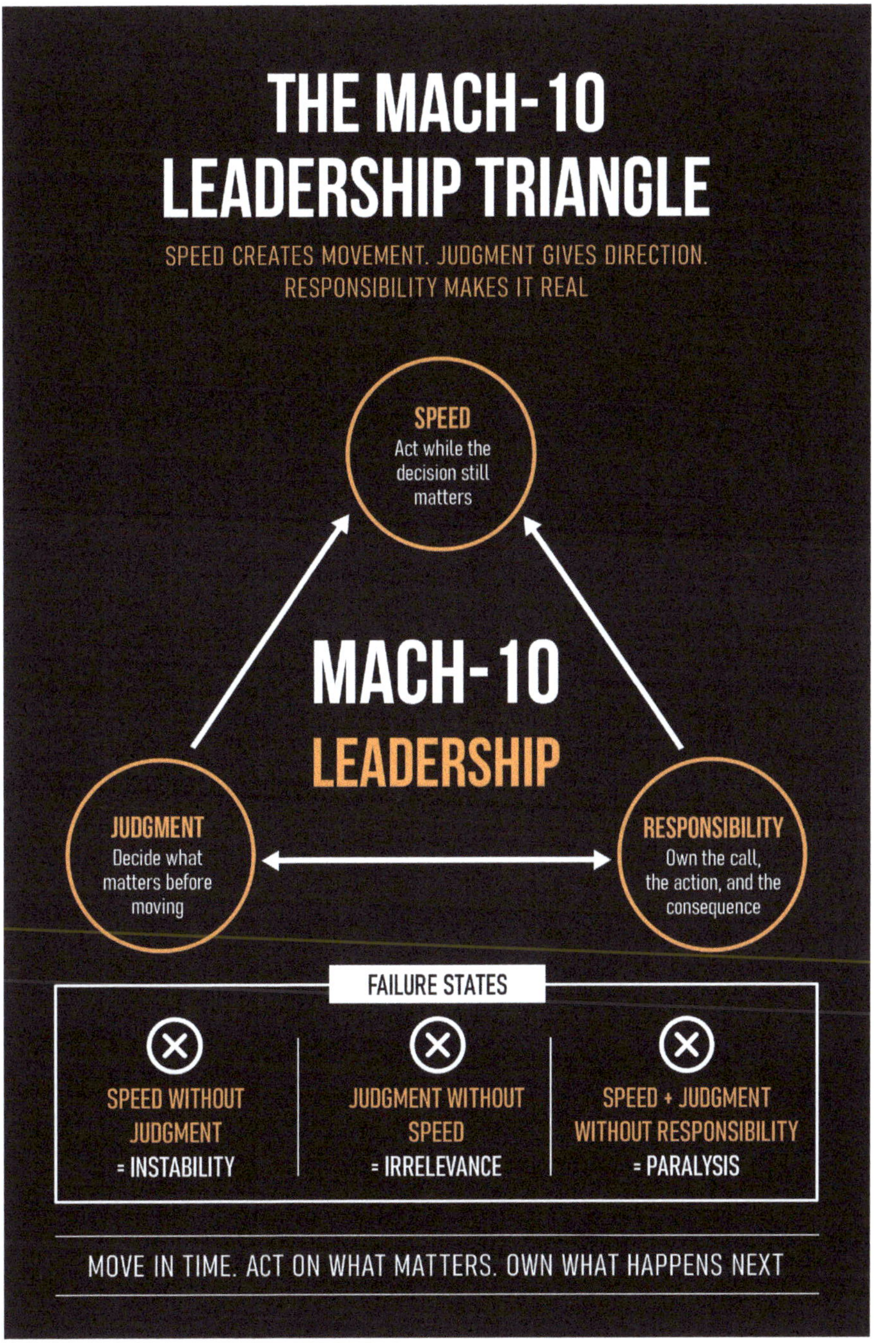

FIG 20.1 – The MACH-10 Leadership Triangle. Speed creates movement, judgment gives direction, and responsibility makes it real.

Each one matters on its own. Together, they determine whether the organization holds under pressure.

The model becomes useful when it is applied to a real decision.

A customer signal appears. Usage is shifting, a competitor is moving, and the team sees the pattern before the organization has decided what to do.

Speed asks how long the decision will retain its value. Judgment asks what matters most and what could create damage if the call is wrong. Responsibility asks who owns the decision, the action, and the correction if the first move misses.

That is the model in motion. All three forces have to stay connected.

The value comes from asking all three questions before the organization defaults to habit. Some teams accelerate before understanding the risk. Others keep analyzing after the window has started closing. Still others reach the right answer and leave it sitting in the room because no one owns what happens next.

That is what turns pressure into leadership.

Speed

Speed is the ability to act while the decision still matters.

It is measured by how quickly the organization moves from signal to action while the window remains open.

When speed is missing, decisions arrive after the moment has passed. Opportunities shift before action happens, and even well-reasoned work loses value because the timing is wrong.

Speed creates the possibility of advantage. It gets the organization through the window before it closes, but it doesn't decide what should move through it.

Judgment

Judgment is the ability to determine what matters before acting.

It tells leaders where to move quickly, where to apply scrutiny, and where more context is required before committing. Without it, teams act on incomplete understanding, chase outputs that look correct but don't fit reality, and confuse activity with progress.

Judgment keeps speed pointed in the right direction. It is the difference between a fast organization and a reckless one.

Responsibility

Responsibility is ownership of the outcome.

Someone has to own the call, the action, and the consequence. Without clear ownership, decisions diffuse, work spreads across teams, alignment expands, and progress slows because no one is accountable for closing the loop.

Responsibility turns decisions into action and keeps speed and judgment from evaporating into discussion.

Every meaningful decision needs an owner.

Not a crowd. Not a vibe. An owner.

Where Systems Break

When these forces fall out of balance, the failures are predictable.

Speed without judgment creates instability. Teams move quickly, but they move on weak understanding. Decisions happen fast while the assumptions beneath them remain thin.

That can look like urgency.

It's often just noise with a shorter deadline.

Judgment without speed creates irrelevance. The thinking improves while the moment disappears. The team gets smarter, the analysis gets cleaner, and the recommendation becomes easier to defend. By the time the decision is ready, the window may already be closing.

That can look like discipline.

It's often delay wearing better shoes.

Speed and judgment without responsibility create drift. The decision may move, but no one fully owns the action, the correction, or what happens next. The room understands the trade-offs. The path is clear enough. Work advances while accountability dissolves.

That can look like empowerment.

It is often abdication with better branding.

Each failure mode has a different surface. Instability looks busy. Irrelevance looks disciplined. Drift looks autonomous.

That's why the failure can hide in plain sight. Everyone can point to something that looks reasonable: speed, analysis, agreement, effort. Plenty is happening, but the decision still lacks what it needs to produce a useful outcome.

The more useful way to read these failures is as diagnostics, not personality flaws. If the same decisions keep arriving late, the answer probably isn't another speech about urgency. If teams keep making fast calls they later unwind, the problem may be weak judgment at the point of action. If good decisions keep dissolving into follow-up meetings, ownership is probably the missing force. The pattern tells you where to intervene.

The Speed with Soul Matrix makes the speed-judgment imbalance visible. Responsibility remains the ownership condition that determines whether movement becomes action.

The matrix isn't a personality test for leadership styles. It's a diagnostic for where force is being lost. The surface symptoms may differ, but the imbalance underneath them is what matters.

The value of the model is that it changes the question. Instead of asking whether the team is moving fast enough, thinking deeply enough, or collaborating broadly enough, leaders can ask whether the decision lacks speed, judgment, or clear ownership.

That makes the diagnosis more precise. A slow decision doesn't always need more urgency. It may need clearer ownership. A reckless decision doesn't always need more process. It may need stronger judgment around the assumptions carrying the most risk. A highly aligned team may still need someone willing to close the loop.

Once the constraint is visible, the response becomes simpler. Leaders can add scrutiny without slowing everything down, assign ownership without reopening the entire discussion, or accelerate the call without pretending the risks have disappeared.

The goal isn't to maximize every dimension. It's to restore what the decision needs to move.

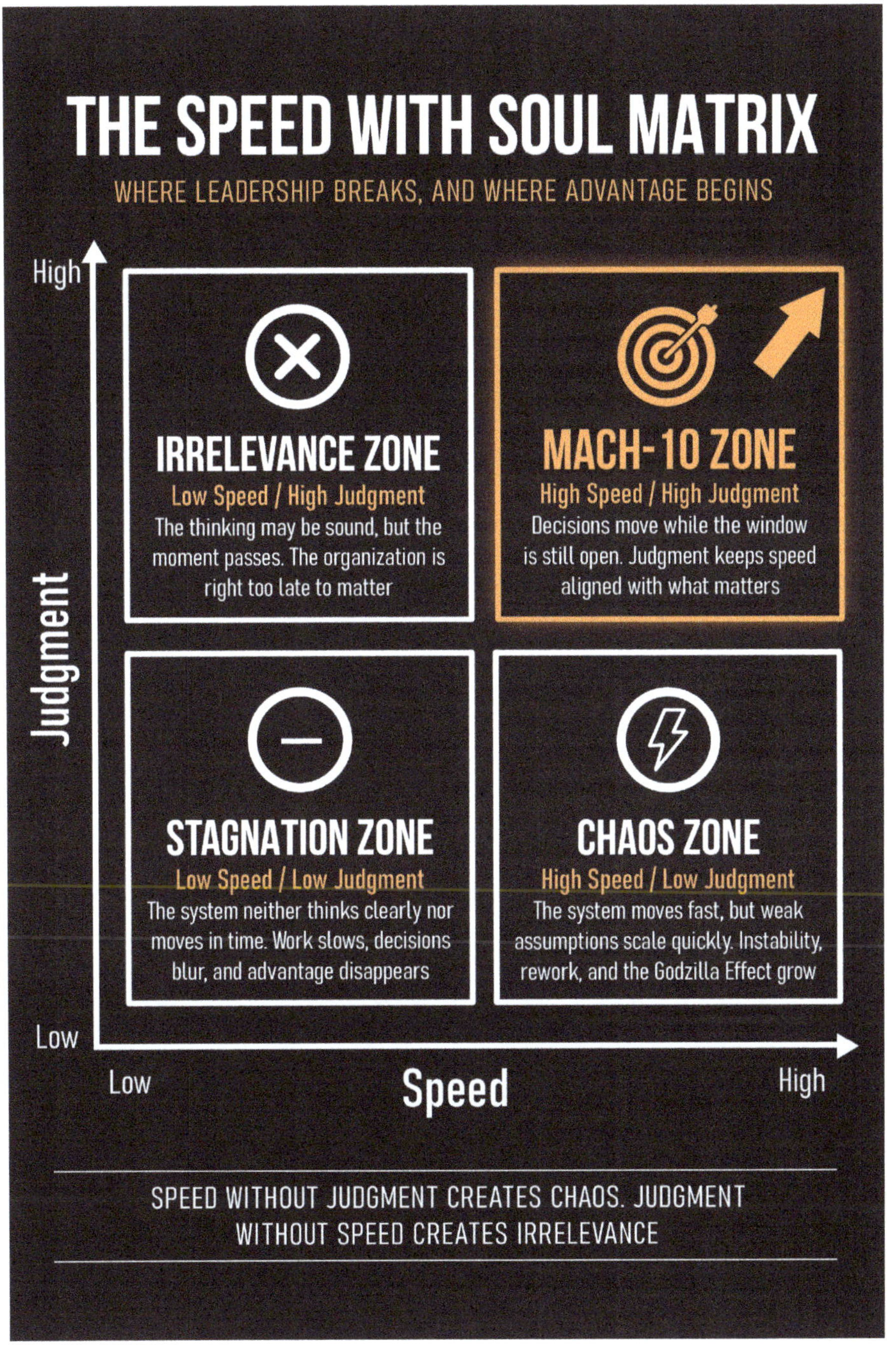
THE SPEED WITH SOUL MATRIX
WHERE LEADERSHIP BREAKS, AND WHERE ADVANTAGE BEGINS
High
Judgment
Low
IRRELEVANCE ZONE
Low Speed / High Judgment
The thinking may be sound, but the moment passes. The organization is right too late to matter
MACH-10 ZONE
High Speed / High Judgment
Decisions move while the window is still open. Judgment keeps speed aligned with what matters
STAGNATION ZONE
Low Speed / Low Judgment
The system neither thinks clearly nor moves in time. Work slows, decisions blur, and advantage disappears
CHAOS ZONE
High Speed / Low Judgment
The system moves fast, but weak assumptions scale quickly. Instability, rework, and the Godzilla Effect grow
Low
Speed
High
SPEED WITHOUT JUDGMENT CREATES CHAOS. JUDGMENT WITHOUT SPEED CREATES IRRELEVANCE

Operating at MACH-10

MACH-10 leaders don't optimize one force at the expense of the others.

They keep the three forces in balance.

They move quickly when timing matters. They apply judgment where the cost of being wrong is high. They assign ownership so decisions turn into action.

They do not wait for perfect conditions.

They operate inside the system as it exists, then keep improving how it moves.

They treat speed, judgment, and responsibility as connected disciplines, not separate virtues.

Over time, this balance compounds.

The system becomes faster and sharper, learning and adjusting in ways that are difficult to replicate.

That's where advantage comes from.

Not from a single decision.

From a system that keeps making better decisions in time to matter.

The Model

Speed determines whether you act in time.

Judgment determines whether you act on what matters.

Responsibility determines whether action actually happens.

Remove any one of them, and the system breaks.

Hold all three, and the system moves.

That's why the model has to stay simple.

Under pressure, leaders don't need another complicated framework to admire.

They need a way to see what's missing before the decision window closes.

That is where the model has to become practical.

Not another ceremony.

Not another checklist people admire and forget.

A way to see what the system needs before the moment passes.

Use the Decision Check when a specific decision starts to slow down, drift, or attract unnecessary coordination.

Use the System Audit when the whole operating system feels slow, noisy, overextended, or harder to steer than it should.

Both tools are built for pressure. They are not meant to make leadership more complicated. They are meant to reveal what is missing before delay, noise, or diffuse ownership starts making the decision for you.

The MACH-10 Decision Check

When a decision starts to slow down, don't add another meeting by reflex. Run the decision through five questions first.

1. What signal are we responding to?

Name what changed. Customer behavior. Market movement. Competitive pressure. Operational strain. Internal risk. If the signal isn't clear, slow down long enough to understand what you're actually reacting to.

2. How long does this decision still matter?

Some decisions hold their value for weeks. Others decay in days, hours, or minutes. If the window is closing, timing is part of quality.

3. What kind of decision is this?

A reversible decision needs movement. A high-consequence decision needs ownership, scrutiny, and a correction path. Treating every decision the same is how teams turn speed into recklessness or delay.

4. Who owns the call?

Input can come from many places. The call can't. If ownership is too broad, the decision will drift.

5. What will tell us if we're wrong?

Every fast decision needs a feedback path. What signal will show the decision is working? What signal will tell you to adjust? What triggers correction?

The goal is to make every decision move at the speed its context requires.

The MACH-10 System Audit

When the system feels slow, noisy, or overextended, look for the missing force.

If speed is missing: Decisions arrive late. Work waits. Signals age before action happens. The team may be careful, but the window closes before the decision can create value.

Ask: Where is motion getting trapped?

If judgment is missing: The system moves, but not always on what matters. Outputs get accepted too easily. Confidence starts replacing understanding. The team moves quickly, but the work drifts away from reality.

Ask: Where are we moving faster than our understanding?

If responsibility is missing: Everyone has input, but no one owns the outcome. Decisions diffuse. Correction slows. Feedback exists, but it doesn't change the next move. The organization stays busy while repeating the same mistakes at higher speed.

Ask: Who owns the call, the action, and the correction?

This audit isn't meant to create another meeting.

It's meant to show where the operating system is losing force.

Once you know what's missing, the next move gets clearer.

Final Principle

You don't need to move perfectly.

You need to move in time, on what matters, with clear ownership.

That is MACH-10 leadership.

The MACH-10 Leader in 90 Seconds

AI compresses time.

When time compresses, decision windows shrink.

When decision windows shrink, latency becomes liability.

A decision only creates value if it arrives while the outcome can still be changed.

Speed alone isn't enough. Speed without judgment creates chaos.

Judgment alone isn't enough. Judgment without speed creates irrelevance.

Responsibility is what turns movement into action, correction, and consequence.

The leader's job isn't to chase speed everywhere. It is to build a system where decisions move at the right speed, judgment enters at the right points, ownership stays clear, and learning compounds faster than delay can destroy value.

That is the MACH-10 Leadership Model.

It isn't a demand to accelerate everything. It's a way to determine what must move, where judgment has to enter, and who owns what happens next.

The model works when speed, judgment, and responsibility remain connected under pressure.

Chapter 21

The Call Is Yours to Make

Leading When the System Moves With or Without You

There comes a point when all of this stops being interesting and starts becoming real.

That point is now.

AI is no longer just a topic, a collection of tools, or a set of frameworks that look impressive in a workshop and then die quietly in a shared drive. It is changing how leadership works when the world around you no longer moves at human planning speed.

The old model gave leaders more time. They could gather input, align the room, refine the plan, and wait until a decision felt safe enough to make.

That world is gone.

Information arrives faster, work advances faster, customers react faster, and markets adjust faster. The tools are absurdly powerful, and they are getting better while you're still deciding whether the team needs another alignment meeting.

That is the strange gift of this moment.

Leaders now have more leverage than ever before. They can synthesize information in minutes, test ideas before the old organization would have finished scheduling the kickoff, generate options, model scenarios, compress research, prototype workflows, and turn raw inputs into action at a speed that used to be impossible.

That leverage becomes dangerous when leadership thinking doesn't evolve with the tools.

Faster tools inside slow leadership do not create advantage. They create a louder version of the same constraint. Work advances, decisions wait, the dashboard improves, and the organization still hesitates.

Somewhere, probably in a meeting called "Strategic Acceleration Sync," everyone wonders why the business doesn't feel faster.

Some leaders will adapt. Many will not.

They will have access to the intelligence, see what is happening, and understand what needs to change. They will still fall behind because changing how you operate is harder than recognizing the problem.

This shift asks more of you. You have to act before everything is fully certain, decide before every stakeholder is perfectly comfortable, and use AI without surrendering judgment. You have to increase pace without becoming reckless and hold ownership when the technology makes it easier to hide behind process, tools, or consensus.

That is the job now.

The Work Is Knowing the Difference

You will feel pressure in both directions.

There will be moments when things start happening faster than you're comfortable with, and your instinct will be to slow them down through more validation, more alignment, more confidence, or one more pass.

That instinct has history behind it. It kept organizations stable for a long time. Follow it too far, though, and it becomes the thing holding you in place while everything else keeps changing.

You will also feel the pull in the other direction. AI will produce answers that look complete. Recommendations will feel clean, and outputs will arrive polished enough to make thinking seem optional.

That pressure is real too. Clean answers are seductive, especially when the clock is running.

Speed without judgment creates instability. Confidence without ownership creates exposure. Automation without understanding simply moves the mess somewhere harder to see.

The work is knowing the difference: when to act, when to pause, when to trust, when to question, when to push authority closer to the work, and when to step in because the cost of being wrong is too high.

That is MACH-10 leadership.

It requires enough confidence to let the organization operate and enough humility to keep testing whether it is heading in the right direction. The leader can't become the permanent brake, but can't disappear behind the pace either. Judgment stays active even when direct involvement becomes selective.

This is neither speed worship nor control theater. It is the discipline to operate at decision speed without losing the human edge.

If you have been in the room long enough, you already know what this feels like.

You have seen the decision that was right and still arrived too late to matter. You have felt the moment when everyone knew what needed to happen and no one closed the loop. You have watched a team wait for certainty while the opportunity quietly left the building, and you have seen a process protect itself long after it stopped protecting the outcome.

Those moments stay with you.

They become scar tissue, improve instinct, and teach you to stop confusing motion with progress.

By this point, the pattern should be visible. Delayed decisions create recovery work, weaker impact, and missed windows. Speed without judgment pushes rework, cleanup, and confusion downstream. AI outputs accepted without interrogation allow small errors to travel before anyone understands them. Success without inspection teaches the organization the wrong lesson.

These are not separate problems. They are the same operating model revealing itself under pressure.

The Moat Is the System

The leaders who figure this out do not wait for perfect conditions. They pay attention to where decisions slow down and fix what surrounds them. They clarify what matters, remove steps that exist only because no one has had the nerve to delete them, and use AI to compress the distance between information and action.

They challenge outputs that look clean but do not hold up. They assign ownership before momentum turns into fog, and they act once there is enough clarity.

Then they learn.

That is where advantage starts compounding.

It doesn't come from one heroic decision, one perfect plan, or one transformation kickoff with a logo and a laminated acronym. It comes from an operating model that keeps producing better decisions while there is still time for those decisions to matter.

That is the shift.

The organization learns, corrects, adapts, and compounds in ways that are difficult to copy from the outside. Competitors can copy a feature, a message, a workflow, or even the tool stack. They can't easily copy leadership habits that consistently produce timely decisions, sound judgment, and clear responsibility.

That is the moat now.

The advantage is built through repetition. One faster decision doesn't change the organization. One clear owner on one project doesn't change the operating model. The shift becomes real when those behaviors become normal enough that people stop treating them as exceptions.

That repetition also changes what competitors are actually chasing. They may copy the visible result and still miss the hundreds of smaller calls, corrections, and trade-offs that produced it. By the time they reproduce what they can see, the organization that created it has already learned something new. Copying the output is easy. Copying the decision rhythm behind it is where things get annoying.

That means the advantage keeps moving too, which is exactly why copying what worked yesterday rarely closes the gap.

Over time, the organization develops a different reflex. Information reaches the people who can act. Teams know which calls belong to them. Leaders engage where judgment adds value instead of where habit tells them to stay involved. Fewer decisions have to fight their way through the company just to reach daylight.

That is when leadership starts compounding. Each clear decision strengthens the next one, each shorter feedback loop improves the next response, and each visible act of ownership makes it harder for ambiguity to return wearing a new meeting title.

Eventually, extraordinary effort is no longer required to create momentum. People act because the priorities are clear, the boundaries are understood, and responsibility no longer has to be rediscovered every time pressure rises. What once required escalation becomes part of the operating rhythm.

You probably already know where your organization is leaking.

Decisions get stuck. Work waits. Alignment becomes a polite substitute for action. AI is used as decoration instead of a working tool, and ownership stays fuzzy while everyone pretends things are moving because everyone is busy.

None of that needs more analysis.

It needs a decision.

The environment will not slow down so you can catch your breath. You can't lead yesterday's organization with tomorrow's tools and expect the math to work.

You have to evolve how you think, how decisions happen, and how responsibility shows up when the pressure arrives.

That doesn't mean blowing everything up. Start where the organization is slow and refuse to let it stay that way.

Choose one workflow, one decision path, or one approval that no longer changes the outcome. Put one clear owner where responsibility used to diffuse. Pull one feedback loop closer to the work. Make one call instead of letting the loop stretch until the advantage disappears.

That is how change becomes real.

It happens through consistent decisions, not one perfect transformation.

You are not here to observe this shift. You are here to lead through it.

If you're still reading, some part of you already knows that.

The organization will keep changing with or without you. The market will keep moving with or without you. The tools will keep accelerating with or without you.

The only question is whether your leadership evolves quickly enough to shape what happens next.

You do not fall behind because of one wrong decision.

You fall behind when too many decisions arrive after they can still matter.

What Holds When Everything Speeds Up

This is the operating truth behind the model.

A decision that arrives too late has no value. Faster learning loops beat slower perfect ones. Alignment has a cost. Spend it intentionally.

Speed without judgment creates chaos. Clarity makes speed useful. Outputs are inputs. Ownership decides.

Trust belongs to the use case, not the tool. AI doesn't eliminate work. It decides where the work shows up.

You do not need perfect information. You need to act in time. Organizations either adapt or lose position.

That's the work now.

Not waiting for permission from the old model. Not asking yesterday's process to approve tomorrow's leadership. Not standing around while the window closes and calling it discipline.

You already know where the organization is slow. You already know where decisions wait. You already know where ownership gets fuzzy, where alignment becomes theater, and where delay keeps sending the invoice.

So make the call. Change the conditions. Own what happens next.

THE SYSTEM WILL
MOVE
THE ONLY QUESTION IS
WHETHER YOU ARE
SHAPING
IT OR REACTING TO IT

About the Author

Jason M. Riggs is a product executive, AI strategist, and author of The MACH-10 PM: AI-Powered Product Management at Hypersonic Speed. He has led product, strategy, revenue, and go-to-market work across companies including GoPro, Qualcomm, PAR Technology, and Audivi AI, where his work has connected technology, customer behavior, execution, and organizational speed.

His work focuses on what happens when AI changes more than the tools teams use. AI changes the tempo of the business itself. Decisions move faster. Work compresses. Signals arrive sooner. The gap between knowing and acting gets smaller. He writes about how leaders can operate inside that pressure without losing judgment, clarity, or accountability.

The MACH-10 Leadership Series grew out of that operating reality. Speed is powerful, but speed without judgment creates noise. AI creates leverage, but leverage without ownership creates risk. The leaders who thrive in this environment do not simply move faster. They build systems that can sense, decide, act, learn, and adjust while the window is still open.

Jason's approach blends practical executive experience with a direct, often irreverent voice for leaders who want useful ideas without the usual beige business-book fog. His writing is built for people operating in real organizations, where decisions are messy, alignment can become theater, and the hardest part of transformation is often changing how the system actually moves.

Before moving into product and technology leadership, Jason worked in radio, which may explain both his bias for rhythm and his strong allergy to corporate bullshit. He lives in Encinitas, California, where he writes, advises, builds, listens to too much music, and keeps trying to make serious leadership ideas feel alive.

Continue the Work

The MACH-10 Leadership Series continues beyond this book with bonus downloadable content, practical toolkits, frameworks, podcast episodes, field notes, and resources built for leaders and product teams working with AI, decision speed, and organizational change.

Visit mach10pm.com to access free resources and subscribe to The MACH-10 Signal, a newsletter on AI-native leadership, product strategy, decision velocity, and building faster systems without losing human judgment.

The work doesn't end here.

The system keeps moving.

Stay sharp.

MACH-10 Leadership Language

A quick reference for the MACH-10 Leadership Model™

Accountable Speed
Moving quickly without letting ownership disappear. Speed only creates value when someone owns the call, the action, and the correction if the decision is wrong.

Automation Trap
The belief that automation removes work when it often moves work somewhere else. Time saved upstream can become review, correction, coordination, or cleanup downstream.

Coordination Tax™
The time and energy consumed by meetings, handoffs, alignment steps, and approval layers before work can create value. Coordination becomes a tax when it slows movement without improving the outcome.

Decision Decay
The loss of decision value as time passes. A decision can be correct and still lose impact if it arrives after the window to act has started closing.

Decision Latency
The time between signal and action. In compressed systems, latency is not just delay. It is value loss.

Decision Value Law™
A decision only creates value when quality, timing, and ownership stay connected. If any one of those elements approaches zero, the decision loses force.

Decision Velocity
The speed at which an organization turns signal into decision, decision into action, action into feedback, and feedback into the next decision.

Godzilla Effect
The organizational damage created when small, plausible AI errors scale through reuse, automation, and defended assumptions before anyone catches them.

Hidden Cost Shift
The downstream work created when automation removes visible effort upstream but pushes review, correction, coordination, or cleanup elsewhere.

Judgment Insertion
The deliberate placement of human judgment earlier in the system, before speed, automation, or AI-generated output compounds a weak assumption.

Learning Loop
The cycle of signal, decision, action, feedback, and adjustment. Advantage compounds when teams shorten that loop without losing judgment, ownership, or connection to reality.

Permission Architecture
The way authority is distributed through the organization. Strong permission architecture places decision rights close enough to the work that teams can move without unnecessary approval.

Speed with Soul™
Moving at decision speed without surrendering judgment, context, ownership, or meaning. Speed creates movement. Soul gives that movement direction.

The MACH-10 Leadership Model™
An operating model for leading in AI-compressed environments: AI compresses time, compressed time shrinks decision windows, shrinking windows turn latency into liability, leaders insert judgment earlier, judgment creates advantage through faster learning loops, and those loops only hold when ownership is clear.

Trust Allocation Model™
The practice of deciding where AI outputs can move freely, where they need friction, and where human ownership must stay close. Trust belongs to the use case, not the tool.

Two-Way Door Audit
A quick decision classification used to determine whether a decision can be reversed with limited cost or requires more scrutiny because the consequences are harder to unwind.

Sources

Selected intellectual anchors behind the MACH-10 Leadership Model

The following works helped shape the intellectual backdrop for the MACH-10 Leadership Model, especially around decision cycles, organizational learning, automation risk, trust calibration, strategy, and judgment under uncertainty.

Bainbridge, Lisanne. "Ironies of Automation." *Automatica*, 1983.

Bezos, Jeff. Amazon shareholder letters, especially discussions of one-way and two-way door decisions.

Boyd, John. Writings and briefings on the OODA loop and decision-cycle advantage.

Brynjolfsson, Erik, and Andrew McAfee. *The Second Machine Age*. W. W. Norton, 2014.

Davenport, Thomas H., and Rajeev Ronanki. "Artificial Intelligence for the Real World." *Harvard Business Review*, 2018.

Edmondson, Amy C. *The Fearless Organization*. Wiley, 2018.

Forsgren, Nicole, Jez Humble, and Gene Kim. *Accelerate*. IT Revolution, 2018.

Kahneman, Daniel. *Thinking, Fast and Slow*. Farrar, Straus and Giroux, 2011.

Klein, Gary. *Sources of Power: How People Make Decisions*. MIT Press, 1998.

Lee, John D., and Katrina A. See. "Trust in Automation: Designing for Appropriate Reliance." *Human Factors*, 2004.

March, James G. "Exploration and Exploitation in Organizational Learning." *Organization Science*, 1991.

Mintzberg, Henry. *The Rise and Fall of Strategic Planning*. Free Press, 1994.

Parasuraman, Raja, and Victor Riley. "Humans and Automation: Use, Misuse, Disuse, Abuse." *Human Factors*, 1997.

Ries, Eric. *The Lean Startup*. Crown Business, 2011.

Rumelt, Richard. *Good Strategy/Bad Strategy*. Crown Business, 2011.

Teece, David J., Gary Pisano, and Amy Shuen. "Dynamic Capabilities and Strategic Management." *Strategic Management Journal*, 1997.